United States Nuclear Regulatory Commission

*Protecting People and the Environment*

NUREG-1948

I0482683

# Final Safety Evaluation Report

## Related to the Aircraft Impact Amendment to the U.S. Advanced Boiling Water Reactor (ABWR) Design Certification

Office of New Reactors

# AVAILABILITY OF REFERENCE MATERIALS
## IN NRC PUBLICATIONS

United States Nuclear Regulatory Commission

*Protecting People and the Environment*

# Final Safety Evaluation Report

## Related to the Aircraft Impact Amendment to the U.S. Advanced Boiling Water Reactor (ABWR) Design Certification

Manuscript Completed:  October 2010
Date Published:  June 2011

Office of New Reactors

# ABSTRACT

This safety evaluation report (SER) documents the technical review of U.S. Advanced Boiling Water Reactor (ABWR) Aircraft Impact Assessment (AIA) application by the U.S. Nuclear Regulatory Commission (NRC) staff. The ABWR AIA application was submitted by the STP Nuclear Operating Company (STPNOC), in accordance with the procedures of Subpart B to Part 52 of Title 10 of the *Code of Federal Regulations* (10 CFR).

The U.S. ABWR is a single-cycle, forced-circulation, boiling water reactor (BWR) with a rated power of 3926 megawatts thermal (MWt) and a design power of 4005 MWt. The staff issued the "Final Safety Evaluation Report (FSER) Related to Certification of the Advanced Boiling Water Reactor Design" (NUREG-1503), on July 13, 1994. The U.S. Nuclear Regulatory Commission issued a final rule certifying the design on May 12, 1997.

On June 30, 2009, STPNOC submitted an application to amend the Design Certification Rule for the ABWR. On September 23, 2010, STPNOC submitted the final amendment application (ML102870017). The purpose of the amendment is to address the requirements of the NRC in 10 CFR 50.150, "Aircraft Impact Assessment." If the proposed amendment is approved, applicants for a combined license (COL) that reference the ABWR standard design may address the requirements of 10 CFR 50.150 by referencing the amended ABWR standard design.

Under 10 CFR 50.150, applicants for new nuclear power reactors are required to perform an assessment of the effects on the designed facility of the impact of a large, commercial aircraft. Using realistic analyses, applicants must identify and incorporate into the design those design features and functional capabilities to show, with the reduced use of operator action, that (1) the reactor core remains cooled or the containment remains intact and (2) spent fuel cooling or spent fuel pool integrity is maintained (referred to as the acceptance criteria). Applicants subject to this regulation are required to submit a description of the design features and functional capabilities identified as a result of the AIA and a description of how those features and capabilities show that the acceptance criteria in 10 CFR 50.150(a)(1) are met with reduced use of operator action.

The NRC determined that the impact of large, commercial aircraft is a beyond-design-basis event. Therefore, applicants may identify either safety-related or non-safety-related features or capabilities to satisfy the requirements for consideration of aircraft impact. The design features relied upon to satisfy the requirements for consideration of aircraft impact may be structures or features (1) whose sole purpose is to address these requirements, or (2) that have a dual purpose of addressing aircraft impact requirements as well as other NRC requirements.

The NRC's review of the applicant's proposed amendment has three objectives. The first objective is to confirm that the applicant has (1) adequately described design features and functional capabilities in accordance with the aircraft impact rule; and (2) conducted an assessment reasonably formulated to identify design features and functional capabilities to show, with reduced use of operator action, that the facility can withstand the effects of an aircraft impact. This evaluation is documented in Chapter 19 of this safety evaluation report (SER).

The second objective of the review is to determine that there will be no adverse impacts from complying with the requirements for consideration of aircraft impacts on conclusions reached by the NRC in its review of the original ABWR design certification (NUREG-1503). This evaluation is documented in Chapters 1, 3, 5, 6, 7, 9, 13, 14, 17, and 20 of this SER.`

The third objective is to determine if the applicant is technically qualified in accordance with 10 CFR 52.47(a)(7) to perform the design work to amend a portion of the ABWR design and to supply the amended portion of the design. This evaluation is documented in Chapter 1 of this SER.

On the basis of its evaluation and independent analyses, the NRC staff concludes that STPNOC's application meets the requirements of Subpart B of 10 CFR Part 52 and 10 CFR 50.150(b). NRC staff issuance of this SER does not constitute a commitment to issue the design certification or in any way affect the authority of the Commission in any rulemaking proceeding pursuant to Subpart H of 10 CFR Part 2 and Subpart B to 10 CFR Part 52.

# CONTENTS

# APPENDICES

# ABBREVIATIONS

| | |
|---|---|
| A2LA | American Association for Laboratory Accreditation |
| ABWR | Advanced Boiling Water Reactor |
| ACLASS | ACLASS Accreditation Services |
| ADS | Automatic depressurization system |
| AIA | Aircraft impact assessment |
| ASME | American Society of Mechanical Engineers |
| ATWS | Anticipated transient without SCRAM |
| COL | Combined License |
| COPS | containment overpressure protection system |
| CUW | Reactor water cleanup system |
| DC | Design certification |
| DCD | Design control document |
| DG | Draft regulatory guide |
| ECCS | Emergency core cooling systems |
| ESP | Early site permit |
| FSAR | Final Safety Analysis Report |
| FSER | Final Safety Evaluation Report |
| GDC | General Design Criterion/Criteria |
| GPM | Gallons per minute |
| HPCF | High-pressure core flooder |
| HPIN | High pressure nitrogen |
| I&C | Instrumentation and control |
| IAS | International Accreditation Service |
| IEEE | Institute of Electrical and Electronic Engineers |
| IEEE Std | IEEE Standard |
| ILAC | International Laboratory Accreditation Cooperation |
| ITAAC | Inspections, tests, analyses, and acceptance criteria |
| ITP | Initial Test Program |
| L-A-B | Laboratory Accreditation Bureau |
| LPM | Liters per minute |
| M&TE | Measuring and test equipment |
| MOV | Motor-operated valve |
| MRA | Mutual Recognition Arrangement |
| NEI | Nuclear Energy Institute |
| NIRMA | Nuclear Information and Records Management Association |
| NIST | National Institute of Standards and Technology |
| NPSH | Net positive suction head |
| NVLAP | National Voluntary Laboratory Accreditation Program |

| | |
|---|---|
| P&ID | Piping and instrumentation diagram |
| QA | Quality Assurance |
| QAPD | QA Program Description |
| RAI | Request for additional information |
| RCIC | Reactor core isolation cooling |
| RCPB | Reactor coolant pressure boundary |
| RG | Regulatory guide |
| RPV | Reactor pressure vessel |
| SBO | Station blackout |
| SCG | Startup Control Group |
| SER | Safety evaluation report |
| SRP | Standard review plan |
| SRV | Safety/relief valve |
| SSC | Structures, systems and component |
| SSE | Safe-shutdown earthquake |
| STPNOC | STP Nuclear Operating Company |
| TS | Technical specification |
| TSD | Technical Support Document |
| WR | Wide range |

# 1.0 INTRODUCTION AND GENERAL DESCRIPTION OF PLANT

## 1.1 Introduction

As part of the application, STP Nuclear Operating Company (STPNOC) proposes changes to Chapter 1. Chapter 1, "Introduction and General Description of Plant," of the Advanced Boiling Water Reactor (ABWR) design control document (DCD) describes the ABWR standard plant design, which includes all buildings primarily dedicated to housing the systems and the equipment related to the nuclear system, or access controls to the equipment and systems. The ABWR design comprises an essentially complete nuclear power plant except for site-specific elements. This section of the safety evaluation report-(SER) discusses the staff's evaluation of the proposed revisions to Chapter 1 and the technical bases for the staff's acceptance of these changes.

## 1.2 Summary of Application

On June 30, 2009, STPNOC (the applicant) submitted an application to amend the Design Certification Rule for the U.S ABWR. That rule approved Revision 4 of the U.S. ABWR DCD. Based on the staff's Requests for Additional Information (RAIs) and the applicant's internal reviews, STPNOC revised this application on May 12, 2010 (U7-C-STP-NRC-1000098, ML101340548), June 17, 2010, (U7-C-STP-NRC-100139, ML101720306), July 12, 2010 (U7-C-STP-NRC-100168, ML102000496), August 4, 2010 (U7-C-STP-NRC-100187, ML102240435), and September 2, 2010 (U7-C-STP-NRC-100202, ML103190120). On September 23, 2010, STPNOC submitted the final application (U7-C-STP-NRC-100213, ML102870017). In these submittals, the applicant proposes the following changes to the certified DCD, Tier 2:

1. Revised Section 1.1.1, "Format and Content," to add a statement to address the location of the information added to comply with Title 10 of the *Code of Federal Regulations* (10 CFR) 50.150.

2. Revised Section 1.1.4, "Design Process," to add a statement to address the design process for STPNOC amendment.

3. Added Subsection 1.2.2.12.23, "Alternate Feedwater Injection (AFI) System," to reference new Section 9.5.14, where the system is summarized.

4. Added Subsection 1.2.2.16.16, "Alternate Feedwater Injection (AFI) Pump House," to describe the location of the AFI Pump House and the equipment housed in the AFI pump house.

5. Revised Figure 1.2-4, "Reactor Building, Arrangement Plan at Elevation 8200 mm" to reflect revisions proposed by the applicant.

6. Revised Figure 1.2-5, "Reactor Building, Arrangement Plan at Elevation 1700 mm" to reflect revisions proposed by the applicant.

7. Revised Figure 1.2-6, "Reactor Building, Arrangement Plan at Elevation 4800/8500 mm" to reflect revisions proposed by the applicant.

8. Revised Figure 1.2-8, "Reactor Building, Arrangement Plan at Elevation 12300 mm" to reflect revisions proposed by the applicant.

9. Revised Figure 1.2-9, "Reactor Building, Arrangement Plan at Elevation 18100 mm" to reflect revisions proposed by the applicant.

10. Revised Figure 1.2-10, "Reactor Building, Arrangement Plan at Elevation 23500 mm" to reflect revisions proposed by the applicant.

11. Revised Figure 1.2-12, "Reactor Building, Arrangement Plan at Elevation 31700/38200 mm" to reflect revisions proposed by the applicant.

12. Revised Section 1.4, "Identification of Agents and Contractors," to add a subsection number (1.4.1) and title to the existing discussion and add new subsections 1.4.2, "Applicant - Aircraft Impact Rule Amendment to Design Certification," and 1.4.3, "Other Contractors and Participants."

13. Revised Section 1.8.3, "Applicability of Experience Information," to add AFI to the list of new and novel design features used in the ABWR.

14. Revised DCD Table 1.9-1, "Summary of ABWR Standard Plant COL License Information," to add the following COL license information items:

| ITEM NO | SUBJECT |
| --- | --- |
| 9.39 | Power Supply for Alternate Feedwater Injection Equipment |
| 9.40 | Test and Surveillance Intervals for Alternate Feedwater Injection Equipment, |
| 19.9k | Procedures for Use of Alternate Feedwater Injection |
| 19.9l | Procedures to Depressurize the Reactor Pressure Vessel (RPV) from the AFI Pump House |
| 19.9m | Verification of Environmental Conditions in the AFI Pump House |
| 19.9n | Description of Electrical Power Supply for AFI Equipment. |

In the DCD amendment dated May 12, 2010 (ML101340548), the applicant deleted the COL license information items listed above.

## 1.3    Regulatory Basis

The regulatory basis and acceptance criteria for reviewing the introduction and general description of plant are in Section 1.0 of NUREG–0800, "Standard Review Plan for the Review of Safety Analysis Reports for Nuclear Power Plants," Revision 3 (SRP).

In particular, the applicable regulatory requirements for the ABWR DCD Amendment for Aircraft Impact Assessment (AIA) are as follows:

- 10 CFR 52.47 requires the technical application information.

- 10 CFR 52.47(a)(7) requires the applicant to demonstrate it is technically qualified to engage in the proposed activities.

## 1.4    Technical Evaluation

NRC staff reviewed the applicant's proposed revision to the certified ABWR DCD Tier 2. The staff reviewed the changes to DCD Chapter 1 and determined that they are necessary and adequately reflect the changes to the certified ABWR design requested by STPNOC.

### 1.4.1 Format and Content (Section 1.1.1)

This section of the ABWR DCD describes the format and content of the DCD. The applicant added a sentence that states, "The STPNOC response to the aircraft impact rule is provided in Tier 2, Appendix 19S." The staff confirmed that this is an accurate description of the information STPNOC is adding to the ABWR DCD and found the applicant's proposed change acceptable.

### 1.4.2 Design Process (Section 1.1.4)

This section of the ABWR DCD describes the process used by the original applicant in designing the ABWR. STPNOC added a sentence that states, "The design process for the STPNOC response to the aircraft impact rule is fully described in the STP 3 and 4 Quality Assurance Program Description, which is provided in Tier 2, Subsection 17.1.19." The staff confirmed that the referenced documents contain an accurate description of the design process used by STPNOC and found the applicant's proposed change to DCD Section 1.1.4 acceptable.

### 1.4.3 General Plant Description (Section 1.2)

This section of the ABWR DCD Section 1.2.2, "Plant Description," provides general descriptions of the plant structures and systems. STPNOC added Subsection 1.2.2.12.23, "Alternate Feedwater Injection (AFI) System," to reference new Section 9.5.14, where the AFI system is described. STPNOC also added Subsection 1.2.2.16.16, "Alternate Feedwater Injection (AFI) Pump House," to describe the location of the AFI Pump House and the equipment housed in the AFI pump house. The staff confirmed that these are appropriate additions to reflect the design changes to the ABWR DCD and found the applicant's proposed changes acceptable.

### 1.4.4 Figures

STPNOC revised several figures in Chapter 1 of the ABWR DCD, as listed in Section 1.2 of this SER, to reflect revisions proposed by the applicant. The staff confirmed that these figures accurately reflect changes to the ABWR DCD and found the applicant's proposed changes acceptable.

### 1.4.5 Identification of Agents and Contractors (Section 1.4)

This section of the ABWR DCD describes the qualifications of the applicant. STPNOC did not propose any changes to this section in its original application. The staff issued RAI 01- 2 requesting STPNOC to provide additional information demonstrating that STPNOC is technically qualified to supply the amended ABWR design in accordance with 10 CFR 52.47(a)(7). The applicant's response to RAI 01-2 dated February 8, 2010 (U7-C-STP-NRC-100037, ML100470589), states that STPNOC is responsible for licensing, operating, maintaining, modifying, decontaminating, and decommissioning STP Units 1 and 2. STPNOC has fulfilled this role since 1997, and has had extensive experience with the design of nuclear structures, systems and components (SSCs). The applicant adds that STPNOC is responsible for the licensing and development of STP Units 3 and 4, including the detailed design of these two planned ABWR units. The applicant further states that its activities as a supplier for the amended portion of the ABWR design certification will be subject to the same controls as those for the STP Units 3 and 4 licensing, design, construction/pre-operation and operations activities affecting the quality and performance of safety-related nuclear plant SSCs and certain activities that are not safety-related but support safe plant operations, or where NRC guidance establishes program requirements.

The applicant stated that STP Units 3 and 4 Quality Assurance (QA) Program provides for control of those activities (as indicated in the DCD Tier 2, Section 17.0) and that appropriate controls ensure compliance with 10 CFR Part 21 during design. The applicant indicated that these requirements are passed down to contractors via procurement documents as appropriate, and vendors are qualified in accordance the QA Program.

The applicant stated that STPNOC has entered into an engineering, procurement and construction (EPC) contract with Toshiba America Nuclear Energy Corporation (TANE), a Delaware corporation, which is a subsidiary of Toshiba. TANE's obligations include supply of the certified ABWR design for STP Units 3 and 4. TANE is the overall Project Manager, and is performing engineering and other responsibilities. TANE is being assisted by Westinghouse Electric (Westinghouse), which is providing engineering for selected systems, primarily fuel and safety analyses and instrumentation and control (I&C) systems. The applicant stated that STPNOC will be responsible as a supplier for the scope of the amendment to comply with 10 CFR 50.150. STPNOC expects to receive support from TANE and Westinghouse under the contracts described above, or under future agreements, as appropriate.

Specifically, with respect to the proposed amendment to the ABWR design certification to comply with the 10 CFR 50.150, the applicant indicated that Westinghouse performed the analyses in accordance with the Nuclear Energy Institute (NEI) 07-13, "Methodology for Performing Aircraft Impact Assessments for New Plant Designs" with assistance from ERIN Engineering & Research, Inc. (ERIN). The applicant stated that ERIN has substantial expertise and familiarity with the methodology described in NEI 07-13, as demonstrated by its significant role on behalf of NEI in the preparation of NEI 07-13. Additionally, Westinghouse has performed this scope of work for another design certification applicant. Sargent & Lundy is responsible for the structural design. The applicant also stated that Sargent & Lundy is an experienced nuclear architect engineer and has extensive experience in the design of structures, systems and components for nuclear power plants.

That applicant stated that, in connection with the EPC contract with TANE, it evaluated the capability of Toshiba to complete the STP Units 3 and 4 ABWR project to NRC standards, and in particular, its qualifications to supply the certified ABWR design in accordance with the requirements of 10 CFR 52.73. STPNOC provided the NRC with a report of the results of that evaluation, describing in detail the bases on which STPNOC determined that Toshiba is qualified to supply the certified ABWR design. Letter from Gregory T. Gibson to NRC re: "South Texas Project, Units 3 and 4, Submittal of Due Diligence Assessment of Toshiba Corporation's Qualification to Supply the Design of the Advanced Boiling Water Reactor," August 19, 2008 (ML082350160). The NRC conducted an inspection to independently assess the basis upon which STPNOC determined that Toshiba is capable of providing the certified ABWR for STP Units 3 and 4. Based on the inspection, the NRC concluded that STPNOC had adequately demonstrated Toshiba's qualification to supply the certified U.S. ABWR for STP Units 3 and 4. Letter from John A. Nakoski to Mark A. McBurnett, STPNOC, re: "NRC Inspection Report 05200012/2009- 202 AND 5200013/2009-202," August 28, 2009 (ML092370709).

Subsequent to submitting its RAI response, STPNOC revised its application and added a subsection number (1.4.1) and title to the existing discussion on the original applicant and add new subsections 1.4.2, "Applicant - Aircraft Impact Rule Amendment to Design Certification," and 1.4.3, "Other Contractors and Participants." The new subsections describe STPNOC's qualifications and those of its primary contractors and subcontractors that supported the amendment to the ABWR design certification. The staff's review of the information added to this section determined that the applicant has adequately described its qualifications and those of its primary contractors and subcontracts and the staff finds the applicant's proposed changes acceptable.

To further evaluate STPNOC's capabilities as a prospective design supplier, the NRC performed an inspection of the applicant's quality controls over the work done by its contractors in support of its request to amend the ABWR design certification to comply with 10 CFR 50.150. The staff's evaluation of information obtained during that inspection is discussed in Section 17.1.19 of this SER.

The staff evaluated the information provided by the applicant in its response to RAI 01-2, its revisions to Chapter 1 of the ABWR DCD, and the results of the staff's inspection of the applicant's quality controls. As stated in the applicant's RAI response, the staff has previously concluded that STPNOC had adequately demonstrated Toshiba's qualification to supply the certified U.S. ABWR for STP Units 3 and 4. Based on the previous determination and the additional information provided by STPNOC in the response to RAI 01-2, the staff determined that STPNOC and its contractors are technically qualified to perform the design work associated with the amended portion of the ABWR design represented by STPNOC's application and to supply the amended portion of the ABWR design. However, based on the staff's evaluation of the information provided in the applicant's RAI response and reviewed during the staff's quality controls inspection the staff determined that STPNOC, by itself, is not technically qualified to supply the amended portion of the ABWR design certification represented in STPNOC's DCD, Revision 3, However, the staff determined that STPNOC and TANE acting together are qualified to supply the amended portion of the ABWR design certification represented in STPNOC's DCD amendment, Revision 3. Therefore, the staff intends to propose inclusion of language to that effect in any rule to certify STPNOC's proposed amendment to the ABWR design certification to ensure that the basis for any NRC finding of technical qualification in support of this design certification amendment remains valid.

1.4.6 Applicability of Experience Information (Section 1.8.3)

This section of the ABWR DCD describes the applicant's evaluation of nuclear field experience. This section also contains a list of the new and novel design features used in the ABWR STPNOC added the AFI system to the list of new and novel design features used in the ABWR. The staff's review determined that this addition accurately reflects the new and novel design features that STPNOC is proposing to add to the ABWR DCD and the staff finds the applicant's proposed changes acceptable.

## 1.5 Conclusion

NRC staff reviewed the general description of the plant design aspects of the STPNOC's application to amend the Design Certification Rule for the ABWR. The staff's review of the changes to DCD Chapter 1 determined that they are necessary and adequately reflect the changes to the certified ABWR design requested by STPNOC. The staff concluded that the applicant has adequately addressed the requirements of 10 CFR 52.47(a)(7). However, the staff's determination is based on the condition that STPNOC and TANE acting together will supply the amended portion of the ABWR design to any future license applicants. The staff will include language to that effect in any rule to certify STPNOC's proposed amendment to the ABWR design certification. The staff reviewed the proposed changes to DCD Tier 2, Chapter 1 and found them acceptable. The staff determined that there would be no adverse impacts from complying with the requirements for consideration of aircraft impacts on conclusions reached by the NRC in its review of the original ABWR design certification (NUREG-1503).

# 3 DESIGN OF STRUCTURES, COMPONENTS, EQUIPMENT, AND SYSTEMS

## 3.2.1 Seismic Classification

### 3.2.1.1 Introduction

As part of the application, STPNOC proposes to add a new AFI system and pump house. This section of the SER discusses the staff's evaluation of the seismic classifications of the proposed additions and the technical bases for the staff's acceptance of these changes. Also, this section discusses the staff's determination of whether the changes adversely impact conclusions reached by the NRC in its review of the original ABWR design certification.

### 3.2.1.2 Summary of Application

DCD Table 3.2-1 and Figure 9.5-6 show the classification of the new AFI system beyond the first check valve from the feedwater system as Non-Safety Class, which is equivalent to non-nuclear safety-related or non-safety-related. Table 3.2-1 also shows both the AFI System and the AFI pump house as Safety Class N and, consistent with the methodology applied in the certified design, the seismic classification of Safety Class N SSCs are identified as not applicable and equivalent to a non-Seismic Category I classification. The amended ABWR DCD Revision 3, Table 3.2-1 shows the instrumentation piping and supports forming part of the containment boundary and a portion of the piping beyond the containment outermost isolation valves as Seismic Category I.

### 3.2.1.3 Regulatory Basis

The regulatory basis and acceptance criteria for reviewing the seismic classification are in Section 3.2.1 of NUREG–0800. The guidance in the Section 3.2.1 of NUREG–0800, "Seismic Classification," references Regulatory Guide (RG) 1.29 for seismic classification of various SSCs. General Design Criterion (GDC) 2, with specific guidance included in RG 1.29 for the seismic classification applies to the extent the changes could impact conclusions that were previously made.

### 3.2.1.4 Technical Evaluation

Staff reviewed the seismic classification of the additional structures, systems and components according to the guidance in SRP 3.2.1. The NRC staff's review determined that the NRC's requirements that apply to the design, construction, testing, operation and maintenance of design features and functional capabilities for design-basis events do not apply to design features or functional capabilities selected by the applicant solely to meet the requirements of the final AIA rule. The staff reviewed the seismic classification of the additional SSCs and determined that the non-safety-related SSCs that are exempt from the seismic classification guidance need not be Seismic Category I, provided that they are evaluated for adverse systems interactions consistent with the certified DCD. As identified in SRP 3.7.2 and SRP 3.12, seismic interactions for structures and piping are evaluated and reviewed in other sections of the DCD and SER. Subsection 3.7.2.8 of the ABWR DCD describes the criteria applied to evaluate the interaction of non-seismic Category I SSCs with Seismic Category I SSCs. DCD Note f to Table 3.2-1 also identifies that equipment that is not safety-related but could damage Seismic

Category I equipment if its structural integrity failed is verified analytically and designed to assure its integrity under seismic loading resulting from the safe-shutdown earthquake (SSE). Therefore, the staff determined that classification of the non-safety-related Safety Class N portion of the AFI system and AFI Pump House as not Seismic Category I is acceptable and consistent with the AIA rule, since the non-safety-related portions of the AFI system and the AFI Pump House are selected solely to meet requirements of the rule and there are criteria established to evaluate adverse systems interactions. The staff's review of the initial Amendment also determined that the nitrogen supply line, nitrogen bottle, and AFI system outside of the pump house were not included in Table 3.2-1 and requested additional information to clarify the classification of these items. NRC staff issued RAI 03.02.02-1 requesting the applicant to review the SSCs being added to the AIA and update classification Table 3.2-1 to include SSCs shown on the piping and instrumentation diagrams.

The applicant's response to RAI 03.02.02-1 dated March 3, 2010, (U7-C-STP-NRC-100056, ML100640162) states that the additional nitrogen gas supply system is not required to meet the AIA rule and the system is being removed from the DCD amendment application. The response adds that the AFI instrument lines and instruments will be included in Table 3.2-1 and states that they are safety-related with the same classification as the existing instrumentation to which it is connected. The amended ABWR DCD, Revision 3 includes these changes and shows the instrumentation piping and supports forming part of the containment boundary as Seismic Category I. Classification of these safety-related items as Seismic Category I is consistent with RG 1.29 and GDC 2 and is therefore acceptable. During its review, the staff also determined that the portion of the AFI system outside of the AFI pump house was not identified in Table 3.2-1. DCD Table 3.2-1 has been revised (ML101190120) to include the non-safety-related AFI piping outside the pump house as non-seismic. Therefore, the staff found the applicant's RAI responses and revised DCD AIA Amendment Table 3.2-1 acceptable. Therefore, RAI 03.02.02-1 is resolved and closed.

Consistent with SRP 3.2.1, electrical items are not within scope of the 3.2.1 review and typically their seismic classification is addressed in Chapter 8. Since Chapter 8 of the application was not revised, the seismic classification of electrical items to support the AFI is evaluated in this subsection. DCD subsection 9.5.14.1 states that the power supply for the pump and motor-operated valves is a non-safety-related power supply and independent of the emergency power supplies and meets the requirements of non-Class 1E power, as described in Chapter 3. Staff concurs that the non-safety-related power supply need not be Class 1E or classified as Seismic Category I and is acceptable in regard to systems interactions, since it is independent of the emergency power supplies.

### 3.2.1.5    Conclusion

The staff reviewed the seismic classification aspects of the STPNOC's application to amend the Design Certification Rule for the ABWR. The staff's review focused on the determination of the appropriate seismic classification for the new AFI system and pump house. The staff found that the applicant has adequately demonstrated that the proposed additions of the AFI pump house, AFI system piping, supports and power supply, as described in the application to amend the ABWR DCD, comply with RG 1.29, GDC 2 and the AIA rule relative to seismic classifications. The staff also found that these seismic classifications do not alter the fundamental safety decisions in the original ABWR design certification (NUREG–1503).

### 3.2.2 Quality Group Classifications

#### 3.2.2.1 Introduction

As part of the application, STPNOC proposes to add a new AFI system and pump house. This section of the safety evaluation discusses the staff's evaluation of the quality group classifications of the proposed additions and the technical basis for the staff's acceptance of these quality group classifications. Also, this section discusses the staff's determination of whether the changes adversely impact conclusions reached by the NRC in its review of the original ABWR design certification.

#### 3.2.2.2 Summary of Application

The amended ABWR DCD Table 3.2-1 and Figure 9.5-6 show the classification of the new AFI System beyond the first check valve from the feedwater system as Non-Safety Class and Safety Class N, respectively, which are equivalent to non-nuclear safety-related or non-safety-related. Noted to Table 3.2-1 identifies that quality group is not applicable to the non-safety-related AFI system and the classification of Safety Class N SSCs in the DCD are typically identified with no quality group classification. In the amended ABWR DCD Revision 3, Table 3.2-1 also shows the instrumentation piping and supports forming part of the containment boundary and a portion of the piping beyond the containment outermost isolation valves as Safety Class 2 and Quality Group B. In the initial Amendment, DCD Figure 6.7-1 showed a modified high-pressure nitrogen gas supply line.

#### 3.2.2.3 Regulatory Basis

The regulatory basis and acceptance criteria for reviewing the quality group classifications are in Section 3.2.2 of NUREG–0800. The guidance in Section 3.2.2 of NUREG–0800, "Quality Group Classification," references RG 1.26 for quality group classification of various SSCs. GDC 1, with specific guidance included in RG 1.26 for the quality group classification applies to the extent the changes could impact conclusions that were previously made.

#### 3.2.2.4 Technical Evaluation

NRC staff reviewed the quality group classification of the additional systems and components according to the guidance in SRP 3.2.2. Consistent with SRP 3.2.2, quality group classification applies to pressure boundary items and their supports but not to structures. The staff's review determined that the NRC requirements that apply to the design, construction, testing, operation and maintenance of design features and functional capabilities for design-basis events do not apply to design features or functional capabilities selected by the applicant solely to meet the requirements of the final AIA rule. The staff determined that SSCs that are exempt from the quality group classification guidance need not be designated with a specific quality group provided that they have appropriate QA requirements applied and are evaluated for adverse system interactions. DCD Note e to Table 3.2-1 identifies that for equipment such as the AFI system that is not safety-related, elements of 10 CFR Part 50, Appendix B are generally applied, commensurate with the importance of the equipment's function. Therefore, the staff finds that appropriate elements of 10 CFR Part 50 are applied to the AFI system and classification of the Safety Class N AFI system with no quality group designation is acceptable and consistent with the certified DCD and the AIA rule. The staff's review also determined that the nitrogen supply line, nitrogen bottle, and AFI system outside of the pump house are not included in Table 3.2-1. In RAI 03.02.02-1, staff requested the applicant to provide additional information to clarify the classification of these items. The staff also asked the applicant to review the SSCs being added in the AIA and to update classification Table 3.2-1 to include the SSCs shown on the piping and instrumentation diagrams (P&IDs).

The applicant's response to RAI 03.02.02-1 dated March 3, 2010 (U7-C-STP-NRC-100056, ML100640162) states that the additional nitrogen gas supply system is not required to meet the AIA rule and the system is being removed from the DCD amendment application. The response further identifies that the AFI instrument lines and instruments will be added to Table 3.2-1 and that they are safety-related with the same classification as the existing instrumentation to which they are connected. The revised DCD Table 3.2-1 designates the AFI instrument lines and instruments as Quality Group B. Classification of these safety-related instrumentation items as Quality Group B is consistent with RG 1.26 and GDC 1 and therefore acceptable. The staff review also determined that the portion of the AFI system outside of the AFI pump house is not addressed and the revised Table 3.2-1 submitted with the RAI response only shows the AFI system located in the AFI pump house. The applicant subsequently revised DCD Table 3.2-1 (ML101190120) to include the non-safety-related AFI piping outside the pump house with no quality group classification, consistent with the methodology in the existing DCD. The applicant also deleted the nitrogen supply modifications. This revised DCD is consistent with the AIA rule, because this non-safety-related portion of the AFI system serves no safety function. Therefore, the staff found the applicant's revised DCD AIA Amendment Table 3.2-1 acceptable, and RAI 03.02.02-1 is resolved and closed.

### 3.2.2.5    Conclusion

The staff reviewed the quality group classifications aspects of the STPNOC's application to amend the Design Certification Rule for the ABWR. The staff's review focused on the determination of the appropriate quality group classification for pressure boundary items and their supports in the AFI system. The staff found that the applicant has adequately demonstrated that the quality group classification of the AFI system, as described in the application to amend the ABWR DCD, comply with RG 1.26, GDC 1 and the AIA rule relative to quality group classifications. The staff also found that these changes do not change the fundamental safety decisions in the original ABWR design certification (NUREG–1503).

### 3.8.6    Alternate Feedwater Injection Pump House

### 3.8.6.1    Introduction

As part of the application, STPNOC proposes to add an AFI pump house to provide enclosure and support to a new AFI system. In addition, STPNOC proposes to add a series of fire doors to Seismic Category I structures. This section of the safety evaluation discusses the staff's evaluation of the proposed additions and the technical basis for the staff's acceptance of these proposed additions. Also, this section discusses the staff's determination of whether the additions adversely impact conclusions reached by the NRC in its review of the original ABWR design certification.

### 3.8.6.2    Summary of Application

The applicant proposes to add or change several DCD sections including DCD Tier 2, Revision 3, Subsections 1.2.2.12.23, "AFI system"; 1.2.2.16.15, "AFI Pump House"; part of Table 1.9-1; part of Table 3.2-1 and the fire doors discussed in DCD Tier 2, Revision 3, Section 9A.4.1.

### 3.8.6.3    Regulatory Basis

The regulatory basis and acceptance criteria for reviewing AFI pump house are in Sections 3.7 and 3.8 of NUREG–0800. In particular, the staff's review is based on 10 CFR Part 50, Appendix A, GDC 2 and 4 and applicable provisions of SRP Sections 3.3.1, 3.3.2, 3.4.2, 3.5.3, 3.7 and 3.8.

### 3.8.6.4    Technical Evaluation

Alternate Feedwater Injection Pump House

The applicant proposes to add a new AFI pump house with a non-safety-related AFI system located outside of the reactor building. Subsection 1.2.2.16.15 of the amended ABWR DCD states that the location of the AFI pump house is a remote distance from the reactor building. The amended Table 3.2-1, "Classification Summary," of the ABWR DCD classifies the AFI pump house as a non-safety class structure; so, the seismic design requirements for Seismic Category I structures in SRP Sections 3.7 and 3.8 for the SSE are not applicable. The staff agreed with the applicant's statement that the seismic design requirements provided in the SRP Sections 3.7 and 3.8 are not applicable to the AFI pump house because of its non-seismic Category I designation.

The staff considered the potential of Seismic II/I interaction effects between the AFI pump house and it's adjacent Seismic Category I structures. Because the AFI pump house will be located at least 300 ft away from Seismic Category I structures, the staff agreed that the seismic II/I interaction issue is not applicable to the AFI pump house.

Additionally, the staff evaluated potential impacts, or changes to the structural and seismic design bases of Seismic Category I SSCs as described in the ABWR DCD that might result from the addition of the AFI pump house. By being located at least 300 ft away from the Seismic Category I structures, the staff concluded that the original structural and seismic design bases of the ABWR Seismic Category I SSCs will remain intact and unaffected by the addition of the AFI pump house. Based on the above evaluation, the staff concluded the addition of the AFI pump house as proposed in the amended ABWR DCD is acceptable.

Replacements and Additions of Fire/Pressure Rated Doors to Seismic Category I Structures

The staff reviewed information related to fire doors presented in Section 9A.4 of the amended ABWR DCD. The review indicated that the applicant proposes to upgrade fire doors as described in Section 9A.4 and the Figures in Sections 1.2 and 9A as part of the fire barriers qualification program. The proposed upgrade involves replacements of nonrated, 3-hour, fire-resistant or non-fire-rated doors with 5-psid door or two 3-hour-rated fire doors for the walls surrounding the above listed fire areas.

Based on the review, the staff issued RAI 03.08.04-4 and requested the applicant to discuss and confirm that the above noted changes and additions of doors as well as their resulting structural configuration changes were evaluated (including wall structural integrity analyses, as needed) to ensure that the changes are bounded within their original structural design basis, and the affected walls will continue to maintain their structural integrity and perform their intended safety functions.

The applicant's response to RAI 03.08.04-4 dated February 25, 2010 (U7-C-STP-NRC-100040, ML100600410) states that the overall structural characteristics of the reactor building are unchanged by replacing these doors. Because the calculations for room pressurization already account for the presence of the doors, the change to two doors or to a pressure-rated door represents a very small change in the analysis. Consequently, replacement of the doors is expected to have a minimal effect on the overall structural performance of the reactor building. The applicant adds that the detailed structural analysis must be performed by the COL Holder following completion of the detailed design which will include such details as reactor building internal wall location, wall dimensions, wall materials, the replacement of the 3-hour fire resistant or non-rated fire doors with 5-psid or two 3-hour-rated fire doors, and any other changes resulting from the final fire hazards analyses. Further, the interior wall structural analysis will be performed as part of the complete reactor building design in accordance with ABWR DCD Tier 1, Section 2.15.10; the associated inspections, tests, analyses, and acceptance criteria (ITAAC) in DCD Tier 1, Table 2.15.10; and Tier 2, Appendix 3H.1.

The staff reviewed the above applicant's response and determined that the response was incomplete and needed to be augmented in order to be acceptable. The staff requested that the applicant provide an ITAAC covering the proposed replacements and additions of fire/pressure rated doors to address the reconciliation of the design basis loads as a result of as-built conditions. The above staff's position was provided to the applicant on May 5, 2010.

In a letter dated May 27, 2010 (U7-C-STP-NRC-100117, Attachment 2, ML101530610), the applicant provided a revised response to RAI 03.08.04-4 stating that:

> The detailed structural analysis must be performed following completion of detailed design, which will include such details as reactor building internal wall location, wall dimensions, wall materials, the replacement of the 3-hour fire resistant or non-rated fire doors with 5-psid or two 3-hour rated fire doors, and any other changes resulting from the final fire hazards analyses. The interior wall structural analysis will be performed as part of the complete reactor building design in accordance with ABWR DCD Tier 1, Subsection 2.15.10, the associated design ITAAC provided in Table 2.15.10, Item 10, and Tier 2, Appendix 3H.1. As identified above, the changes and additions of fire doors, as identified in the Appendix 9A markups for the DCD amendment application, are within the scope of these existing analysis requirements. These requirements ensure that the changes and additions of fire doors are considered in the structural design basis and evaluation of the design basis loads in accordance with ABWR DCD Tier 1, Subsection 2.15.10.

The staff evaluated the above revised response to RAI 03.08.04-4 and concluded that the replacements and additions of fire/pressure rated doors will be included in the original structural design basis as referenced in DCD Tier 1, Section 2.15.10, and the implementation of the associated ITAAC design commitments within DCD Tier 1, Table 2.15.10, Item 10. Therefore, the staff has reasonable assurance that the certified design will not be negatively impacted by replacements and additions of the fire/pressure rated doors to Seismic Category I structures. Based on this finding, RAI 03.08.04-4 is considered resolved and closed.

<u>Impact of the Addition of the AFI System on the Adequacy of the Seismic Design Basis of the</u>
<u>Original Feedwater Piping and Supports</u>

The staff reviewed the information related to the proposed addition of the AFI system, including the amended ABWR DCD Figure 9.5-6, "Alternate Feedwater Injection System Schematic." The staff's review assessed the potential impact on the adequacy of the seismic design basis of the original feedwater piping and supports. Based on the review, the staff issued RAI 03.08.04-2 requesting that the applicant confirm whether a piping seismic response analysis of the modified reactor water cleanup system (CUW) tie-in lines to the feedwater system, modeled in conjunction with the AFI line, was performed to ensure that the effects of the AFI line addition are accounted for within the seismic design basis of the original feedwater piping and supports. Also, the staff asked the applicant to clarify whether the design basis of the feedwater system will be adversely affected by the addition of the AFI line. Finally, the staff asked the applicant to summarize the results of these analyses, if any.

The applicant's response to RAI 03.08.04-2 dated February 25, 2010 (ML100600410) states that the piping seismic response analysis of the modified CUW tie-in line to the feedwater system has not yet been performed. The applicant adds that this analysis must be performed by the COL Holder following completion of detailed design and after determination of such details such as pipe routing, location of pipe supports and restraints, and final line sizes. In addition, the final piping seismic analysis of the feedwater system will account for the effects of attached non-safety-related piping systems, including the CUW water system and the AFI piping. Furthermore, this piping seismic analysis will be performed according to the requirements identified in the ABWR DCD Tier, Section 3.3, the associated ITAAC provided in Table 3.3, and the seismic analysis methods identified in Tier 2, Section 3.7.3. Because the tie-in of the AFI line is to the non-safety-related portion of the CUW lines, and consequently does not tie-in directly to a safety-related line, the applicant expects that the design basis of the original feedwater piping and supports will not be adversely affected by this added AFI tie-in line.

The staff reviewed the above applicant's response to RAI 03.08.04-2 and determined that the response was incomplete and needed to be augmented in order to be acceptable. The staff requested that the applicant provide an ITAAC covering the proposed AFI tie-in line to address the reconciliation of the design basis loads as a result of the as-built conditions. The above staff's position was provided to the applicant on May 5, 2010.

In a letter dated May 27, 2010 (U7-C-STP-NRC-100117, Attachment 1, ML101530610), the applicant provided a revised response to RAI 03.08.04-2 stating that:

> The piping seismic response analysis of the modified reactor water cleanup system (CUW) tie-in lines to the feedwater system has not yet been performed. This analysis must be performed following completion of detailed design after determination of such details as pipe routing, location of pipe supports and restraints, final line sizes, etc. The final piping seismic analysis of the feedwater system will account for the effects of attached non-safety piping systems, including the CUW water system and the AFI piping. This piping seismic analysis is performed according to the requirements identified in the ABWR DCD Tier 1, Section 3.3, the associated design ITAAC provided in Table 3.3, Items 1 through 3, and the seismic analysis methods identified in Tier 2, Section 3.7.3. As identified above, adding AFI piping and tie-in to the CUW system is within the scope of these existing analysis requirements for the piping design ITAAC.

These requirements ensure that any effects of the non-safety-related AFI piping on ASME Class 1, 2 and 3 piping are considered. Because the tie-in of the AFI line is to the non-safety-related portion of the CUW lines, and consequently does not tie-in directly to a safety-related line, it is expected that the design basis of the original feedwater piping and supports will not be adversely affected by this added AFI tie-in line.

The staff evaluated the above revised response to RAI 03.08.04-2. The staff verified that the plant-specific piping layout and support configuration data that show the tie-in of the AFI line is to the non-safety-related portion of the CUW lines and does not tie-in directly to a safety-related line. The staff concluded that the piping seismic response analysis of the modified CUW tie-in lines to the feedwater system will be accounted for within the seismic design basis of the original feedwater piping and supports. Therefore, the staff has reasonable assurance that the certified design will not be negatively impacted by the addition of the AFI tie-in line. Based on this finding, RAI 03.08.04-2 is considered resolved and closed.

### 3.8.6.5    Conclusion

NRC staff reviewed the structural and seismic design aspects of the STPNOC's application to amend the Design Certification Rule for the ABWR. Specifically, the applicant proposes to add an AFI pump house, a new AFI System, a series of fire doors, and other related components. The staff's review focused on the determination of structural and seismic safety and design adequacy of the (1) proposed addition of an AFI pump house, (2) proposed replacements and additions of fire- and pressure-rated doors to Seismic Category I structures, and (3) impact assessment of the addition of the AFI System on the adequacy of the seismic design basis of the original feedwater piping and supports. The staff determined that there would be no adverse impacts from complying with the requirements for consideration of aircraft impacts on conclusions reached by the NRC in its review of the original ABWR design certification (NUREG-1503).

The applicant also provided reasonable assurance that the original structural and seismic design bases included in the ABWR DCD for Seismic Category I SSCs will remain intact and unaffected by the addition of the AFI pump house and replacements and additions of fire/pressure rated doors to Seismic Category I SSCs.

The staff also found that the proposed modifications do not change the fundamental safety decisions in the original FSER. Therefore, the structural and seismic aspects of the proposed amendment to the DCD are acceptable.

# 5 REACTOR COOLANT SYSTEM AND CONNECTED SYSTEMS

As part of the application, STPNOC revises DCD Figure 5.1-3, "Nuclear Boiler System P & ID, (Sheet 4 of 11)," to show the AFI system connection to the Nuclear Boiler System through the non-safety-related portion of the CUW tie-in lines to the feedwater system. The staff found this change acceptable, and is evaluated in Section 9.5.14 of this SER.

## 5.2.2    Overpressure Protection

### 5.2.2.1    Introduction

As part of the application, STPNOC proposed to add a solenoid valve to a SRV for overpressure protection. The overpressure protection in the ABWR is provided using 18 SRVs, of which 8 are part of the pressure set point groups and mounted on the four main steam lines between the reactor vessel and the first isolation valve inside the containment. This section of the SER safety evaluation discusses the staff's evaluation of the over pressure protection of the proposed changes and the technical basis for the staff's acceptance of these proposed changes. Also, this section discusses the staff's determination of whether the changes adversely impact conclusions reached by the NRC in its review of the original ABWR design certification.

### 5.2.2.2    Summary of Application

In the initial submittal dated June 30, 2009 (ML092040048), the applicant proposed to add a paragraph to DCD Subsection 5.2.2.4.1 describing the addition of a solenoid valve to SRV–E and a nitrogen supply line to supply nitrogen from the AFI pump house. The intent of this modification was to open one SRV from the pump house after the aircraft impact. The applicant also proposed changes to the DCD Figures 5.1-3 "Nuclear Boiler System P & I D," (Sheets 1, 2, 9 and 11 of 11) to incorporate the new solenoid valve for the mitigating functions from the pump house.

However, in response to RAI 06.02.04-1, the applicant, in letter dated January 13, 2010 (U7-C-STP-NRC-100009, ML100190088), withdrew the changes in the DCD Section 5. Therefore, there are no changes to the overpressure protection system, including the SRVs, due to the AIA Amendment. Also, in response to RAI 05.02.02-1, the applicant reiterates and references the response to RAI 06.02.04-1. Therefore, there are no changes to the overpressure protection system, including the SRVs, in regard to the AIA Amendment. In the revised DCD AIA Amendment dated May 12, 2010 (ML101340548), the applicant withdrew the changes.

### 5.2.2.3    Regulatory Basis

The regulatory basis for reviewing the overpressure protection is in Section 5.2.2 of NUREG-0800, Revision 3.

In particular, the acceptance criteria are based on meeting 10 CFR Part 50, Appendix A, GDC 15 and 31. The acceptance criteria are based on GDC 15, as it relates to the reactor coolant system and associated auxiliary, control, and protection systems being designed with sufficient margin to ensure that the design conditions of the Reactor Coolant Pressure Boundary (RCPB) are not exceeded during any condition of normal operation including anticipated operating occurrences (AOOs). Because the amendment included an addition of a new solenoid valve to modify the operation of SRV-E, it is required that the safety valves have sufficient capacity to perform their intended safety-related function of limiting the pressure to less than 110 percent of the RCPB design pressure.

SRP Section 5.2.2 states that the acceptance criteria are based on GDC 31, as it relates to the fracture behavior of the RCPB. Overpressure protection during low temperature operation is not considered for the ABWR or addressed because there is a very low probability that the ABWR will operate in water-solid conditions. Therefore, overpressure protection during low temperature conditions is not addressed for the ABWR.

### 5.2.2.4    Technical Evaluation

NRC staff reviewed the information in the DCD amendment. The staff issued RAI 05.02.02-1 requesting the applicant to confirm that the addition of the new solenoid valve will not prevent the SRV-E from performing its intended safety-related function. The applicant's response to RAI 05.02.02-1 dated January 20, 2010 (U7-C-STP-NRC-100026, ML100250139), states that the additional nitrogen gas supply system and the additional solenoid valve described in the initial DCD amendment submittal is not required to meet the AIA rule. The overpressure protection system is not affected by any of the design features incorporated into the design to meet the AIA rule. The SRVs will open and close as the reactor pressure increases and decreases. Emergency core cooling systems (ECCS) and the ADS will be available for core cooling. Core cooling is achieved without any modifications to the overpressure protection and ECCS. The staff's evaluation of core cooling is described in SER Subsection 19.S.4.2, "Key Design Features for Core Cooling" and Subsection 19.S.4.3, "Key Design Features that Protect Core Cooling Design Features" of this SER. The staff reviewed the revised DCD AIA Amendment and confirmed that there are no changes to the certified DCD Section 5.2.2. Therefore, RAI 05.02.02.-1 is resolved and closed.

In addition, the staff issued RAI 05.02.02-2 requesting the applicant to describe the type of analyses performed to ensure that the non-safety AFI line connected to the CUW tie-in lines to the feedwater system is designed so that it will not inadvertently impact the ability of safety-related RCIC system, which injects water into the reactor through the feedwater system, to perform its intended functions. This RAI is addressed in Subsection 9.5.14.4 of this SER because it directly relates to the AFI system.

### 5.2.2.5    Conclusion

NRC staff reviewed the overpressure protection aspects of the STPNOC's application to amend the Design Certification Rule for the ABWR. The staff's review found the applicant has adequately addressed the applicable requirements. The staff determined that there would be no adverse impacts from complying with the requirements for consideration of aircraft impacts on conclusions reached by the NRC in its review of the original ABWR design certification (NUREG-1503).

# 6 ENGINEERED SAFETY FEATURES

## 6.2.4    Containment Isolation System

### 6.2.4.1    Introduction

As part of the application, STPNOC proposes changes to the containment isolation information for high-pressure nitrogen gas supply system.  This section of the ABWR DCD addresses the isolation systems including valves and associated piping that allow normal or emergency passage of fluids through the containment boundary while preserving the ability of the boundary to prevent or limit the escape of fission products from postulated accidents.  This section of the SER discusses the staff's evaluation of the containment isolation system of the proposed additions and the technical basis for the staff's acceptance of these proposed changes.  Also, this section discusses the staff's determination of whether the changes adversely impact conclusions reached by the NRC in its review of the original ABWR design certification.

### 6.2.4.2    Summary of Application

In the initial submittal dated June 30, 2009 (ML092040048), the applicant proposed changes to DCD Section 6.7, "High Pressure Nitrogen Gas Supply System," and to Section 6.2.4, "Containment Isolation System."  The applicant provided markups to the DCD to add an additional non-safety-related nitrogen gas storage bottle capable of supplying nitrogen to one of the non-ADS SRVs from the AFI pump house to allow system depressurization in the event of the loss of nitrogen supply in the reactor building.  In the revised DCD AIA Amendment dated May 12, 2010 (ML101340548), the applicant removes the proposed changes to DCD Section 6.7.

### 6.2.4.3    Regulatory Basis

The regulatory basis for reviewing containment isolation system is in Section 6.2.4 of NUREG-0800, Revision 3.

### 6.2.4.4    Technical Evaluation

The staff reviewed the proposed changes and issued RAI 06.02.04-1 requesting the applicant to provide additional details on the design and operation of added containment isolation valves P54-F301/F302 shown in Figure 6.7-1 and Table 6.2-7 and associated other changes in the DCD amendment markups.  In response to this RAI in the letter dated January 13, 2010 (U7-C-STP-NRC-100009, ML100190088), the applicant states that because the additional nitrogen gas supply system described in the initial DCD amendment submittal is not required to meet the AIA Amendment rule, this system is being removed from the DCD amendment.  Consequently, all of the DCD amendment markups associated with this system will be deleted.  As a result, there will no longer be a change to DCD Figure 6.7-1 and Table 6.2-7 and associated other changes of the DCD for this amendment.  On May 12, 2010, the applicant submitted a revised DCD AIA amendment (ML101340548).  The staff reviewed the revised DCD AIA Amendment, and confirmed that the applicant is not modifying the containment isolation system.  Based on the applicant's revised amendment, RAI 06.02-04-1 is considered resolved and closed.

### 6.2.4.5 Conclusion

NRC staff reviewed the containment isolation system of the STPNOC's application to amend the Design Certification Rule for the ABWR. The staff's review concluded that the containment isolation system is not modified to comply with the AIA rule and remains unchanged as certified originally in NUREG–1503.

# 7 INSTRUMENTATION AND CONTROLS

## 7.7    Instrumentation and Controls not Required for Safety

### 7.7.1    Introduction

As part of the application, STPNOC proposes to add additional instrumentation (reactor water level, suppression pool water level, wetwell pressure, and reactor vessel pressure) in the AFI pump house. The information in this chapter emphasizes those instruments and associated equipment that are applicable to the ABWR design amendment of the ABWR DCD that addresses the AIA rule. This section of the SER discusses the staff's evaluation of the I&C required for the proposed additions and the technical basis for the staff's acceptance of these proposed changes. Also, this section discusses the staff's determination of whether the changes adversely impact conclusions reached by the NRC in its review of the original ABWR design certification.

### 7.7.2    Summary of Application

Assessment of the AIA rule on the ABWR reactor design was performed by the applicant in accordance with 10 CFR 50.150(a) to identify and incorporate into the design those design features and functional capabilities to show that with the reduced use of operator actions, (i) the reactor core remains cooled or the containment remains intact; and (ii) spent fuel cooling or spent fuel pool integrity is maintained. In Section 7.7, "Control Systems Not Required for Safety," the DCD amendment addresses additional features of the design to comply with the NRC AIA rule.

The applicant identifies a need for an AFI system that is included to conform to the AIA rule and is located in a remote facility. Section 7.7 of the DCD amendment references Section 9.5.14, which contains the details of the AFI system I&C. The AFI system is designed to mitigate the consequences of an aircraft impact.

The AFI system is capable of injecting water (≥3028 Lpm [800 gpm]) into the RPV at operating pressure and is located outside of the reactor building. The system is designed to be capable of providing sufficient core cooling in the unlikely event that all normal and emergency core cooling systems are unavailable. The power supply for the AFI system is non-safety-related and is independent and physically separated from the emergency power supplies such that a simultaneous loss due to a beyond design basis event is unlikely.

The following I&C signals are provided in the AFI pump house:

- RPV water level
- RPV pressure
- Wetwell wide range (WR) pressure
- Suppression pool water level
- AFI pump flow and discharge pressure
- AFI dedicated water storage tank water level

### 7.7.3    Regulatory Basis

The regulatory basis and acceptance criteria for reviewing the instrumentation and controls not required for safety are in Section 7.7 of NUREG–0800.

In particular, the following acceptance criteria are applicable to this review:

a.    10 CFR 50.55a(h), "Protection and safety systems," requires compliance with IEEE Std 603-1991, "IEEE Standard Criteria for Safety Systems for Nuclear Power Generating Stations," and the correction sheet dated January 30, 1995.  For control systems isolated from safety systems, the applicable requirements of 10 CFR 50.55a(h) are defined in IEEE Std 603-1991 Clause 5.6.3, "Independence Between Safety Systems and Other Systems," and IEEE Std 603-1991 Clause 6.3, "Interaction Between the Sense and Command Features and Other Systems."

b.    GDC 24, "Separation of protection and control systems," states that the protection system shall be separated from control systems to the extent that failure of any single control system component or channel, or failure or removal from service of any single protection system component or channel which is common to the control and protection systems leaves intact a system satisfying all reliability, redundancy, and independence requirements of the protection system.  Interconnection of the protection and control systems shall be limited so as to ensure that safety is not significantly impaired.

c.    10 CFR 52.47(b)(1) requires that a design certification (DC) application contain the proposed inspections, tests, analyses, and acceptance criteria (ITAAC) that are necessary and sufficient to provide reasonable assurance that, if the inspections, tests, and analyses are performed and the acceptance criteria met, a plant that incorporates the design certification has been constructed and will operate in accordance with the design certification, the provisions of the Atomic Energy Act, and the NRC's regulations.

### 7.7.4    Technical Evaluation

The purpose of the staff's evaluation is to assess the impact of the AFI system's I&C on the plant's existing control and protection systems.  The AFI system's I&C is independent and separate from the normal plant control and protection systems.  However, the field sensors associated with the AFI system share process connections with safety-related plant I&C sensors.  The staff's evaluation is to verify that the normal plant safety and non-safety-related I&C systems are not adversely impacted by the interfacing AFI system's I&C.

10 CFR 50.55a(h), which requires compliance with IEEE Std 603-1991, and GDC 24 require a non-safety-related system should be designed so as to not prevent a safety system from performing its intended function and to prevent the effects of a single random failure.  The staff asked the applicant to demonstrate that the I&C for the AFI system does not adversely impact the plant's existing safety-related and important-to-safety I&C systems.  In response to RAI 07.07-1, dated March 3, 2010 (U7-C-STP-NRC-100056, ML100640162), the applicant states that there are no automatic controls or functions associated with the AFI system, and it's I&C are hard-wired and manually initiated.  In addition, the applicant states that the AFI system's I&C is only used for indication purposes, and the water level and pressure instrumentation in the AFI pump room use a separate set of transmitters and a separate power supply independent of the existing I&C.  As part of the response to RAI 07.07-1, the applicant also provides changes to the AIA DCD amendment application to clarify the safety classification

of the I&C components in the AFI system, which the staff found acceptable. The applicant's response to RAI 07.07-1, which references RAI 03.02.02-1, states that the safety classification of the AFI I&C sensors and piping is the same as that of the existing interfacing I&C system. The applicant's response to RAI 03.02.02-1 provides a markup of DCD Tier 2, Table 3.2-1, "Classification Summary," which now includes the AFI system's I&C and associated piping with Seismic Category I classification. Therefore, the staff found that the Seismic Category I AFI transmitters that share the safety-related process connections will not adversely impact the plant's existing safety functions. The staff found that the DCD amendment complies with the acceptance criteria for 10 CFR 50.55a(h) and GDC 24. The staff reviewed the revised DCD AIA Amendment and confirmed that the proposed changes are incorporated. Therefore, RAI 07.07-1 and RAI 03.02.02-1 are resolved and closed.

The staff asked the applicant to provide an ITAAC in the DCD amendment to demonstrate that the I&C included in the AFI system will not adversely affect and is adequately isolated from the plant safety systems. The applicant's response to RAI 07.07-3 dated April 8, 2010 (U7-C-STP-NRC-100049, ML101040254), states that the safety classification of the I&C and piping for the AFI system is the same as that for the existing plant I&C to which it is connected. The applicant's response to RAI 03.02.02-1 provides a markup of DCD Tier 2, Table 3.2-1, "Classification Summary," which includes the AFI-related I&C classified as Seismic Category I. The applicant's response to RAI 07.07-3 and RAI 14.02-1 dated April 8, 2010 (U7-C-STP-NRC-100049, ML101040254), provides a markup of the ABWR DCD Tier 1 that includes an ITAAC for the AFI system. ITAAC Table 2.11.24, Item 12 includes a design commitment to verify that an AFI instrumentation device that is physically attached to instrumentation piping satisfies the same requirements (safety class, quality group, and seismic category) as the instrumentation piping to which it is attached. Also in this response, the applicant references an existing ITAAC (Table 3.3, Item 1) verifying that the AFI instrumentation piping meets the quality requirements identified in Tier 2, Table 3.2-1. The staff found that these ITAAC are adequate to verify that the AFI system's I&C will not have an adverse impact and is adequately isolated from the plant's existing control and protection systems, and therefore are in compliance with 10 CFR 52.47(b)(1). The staff reviewed the revised DCD AIA Amendment and confirmed that the proposed changes are incorporated. Therefore, RAI 03.02.02-1, RAI 07.07-3 and RAI 14.02-1 are resolved and closed.

### 7.7.5   Conclusion

NRC staff reviewed the AFI system's I&C of the STPNOC's application to amend the Design Certification Rule for the ABWR. The staff performed an evaluation to confirm that the AFI system's I&C interface with the plant's safety and control systems conforms to the applicable regulations and does not adversely impact the plant's existing safety and control functions. The staff used applicable criteria discussed in Section 7.7.3 of this SER to evaluate the AFI system's I&C. The staff found that the AFI system's I&C described in the DCD amendment satisfies the applicable acceptance criteria of SRP Section 7.7, which address 10 CFR 50.55a(h), 10 CFR 52.47(b)(1), and GDC 24, and does not adversely impact the plant's existing safety and control systems. The staff determined that there would be no adverse impacts from complying with the requirements for consideration of aircraft impacts on conclusions reached by the NRC in its review of the original ABWR design certification (NUREG-1503).

# 9  AUXILIARY SYSTEMS

## 9.5.1     Fire Protection Program

### 9.5.1.1     Introduction

This section describes the NRC staff's evaluation of the changes to the credited fire protection program within the certified ABWR design.  This section of the safety evaluation discusses the staff's evaluation of fire protection with the proposed changes and the technical basis for the staff's acceptance of these proposed changes.  This section also discusses the staff's determination of whether the changes adversely impact conclusions reached by the NRC in its review of the original ABWR design certification.

### 9.5.1.2     Summary of Application

The applicant has submitted a design certification amendment incorporating several fire protection program changes.  These fire protection-related design changes are reported in the amended ABWR DCD Tier 2, Revision 3, Appendix 19S, with further details described in Section 9.5.1 and Appendix 9A that include Figures 9A-1 through 9A-11.  These changes include the following:

- a non-fire-rated floor becoming a rated 3-hour fire barrier
- addition of fire-rated floor plugs
- several regular fire doors becoming capable of withstanding a 5 psid
- all fire barriers, including associated penetration seals, within the reactor building capable of withstanding 5 psid
- several fire areas split into two or more fire areas

### 9.5.1.3     Regulatory Basis

The regulatory basis for reviewing the fire protection program is in Section 9.5.1 of NUREG-0800, Revision 5.  The relevant requirements are: 10 CFR 50.48 and 10 CFR Part 50, Appendix A, GDC 3, and guidance of RG 1.189, Revision 1.

### 9.5.1.4     Technical Evaluation

NRC staff reviewed the description of certified design changes relating to fire protection.  The proposed changes from a non-rated barrier to a fire rated barrier or the splitting of a fire area into smaller fire areas increases the compartmentation, reduces the spread of fire, and increases protection of safe shutdown equipment.  Changing a regular 3-hour fire barrier to a barrier with the same fire rating plus an increased pressure capability reduces the spread of fire and increases the protection of safe shutdown equipment.  The applicant's changes do not invalidate or reduce the fire protection capabilities of the certified design, but rather increase the protection worth.

### 9.5.1.5 Conclusion

NRC staff reviewed the fire protection system of the STPNOC's application to amend the Design Certification Rule for the ABWR. NRC staff's review concluded that the applicant has adequately addressed the requirements of 10 CFR 50.48 and 10 CFR Part 50, Appendix A, GDC 3. The staff determined that there would be no adverse impacts from the amended design on conclusions reached by the NRC in its review of the original ABWR design certification (NUREG-1503). Therefore, staff found these changes acceptable.

### 9.5.14 Alternate Feedwater Injection System

### 9.5.14.1 Introduction

As part of the application, STPNOC proposes to add an AFI system. The AFI system is a reactor coolant makeup system capable of providing sufficient core cooling in the unlikely event that all normal and emergency core cooling systems are unavailable. The system has a dedicated water source and injection is provided through the non-safety-related portion of the CUW tie-in lines to the feedwater system. The AFI system capacity and discharge pressure (at rated pressure) are comparable to the same parameters in the high-pressure core flooder (HPCF) system. This section of the safety evaluation discusses the staff's evaluation of the proposed additions and the technical basis for the staff's acceptance of these proposed changes. Also, this section discusses the staff's determination of whether the changes adversely impact conclusions reached by the NRC in its review of the original ABWR design certification.

### 9.5.14.2 Summary of Application

DCD Revision 3, Tier 2, Section 9.5.14, "Alternate Feedwater Injection System," describes the AFI system. The system is capable of injecting into the RPV at operating pressure and providing sufficient core cooling in the event that normal and emergency cooling systems are not available.

A schematic of the system is depicted in DCD Tier 2, Figure 9.5-6. The system is in the AFI pump house. The AFI system takes suction from a water source located near the AFI pump house and discharges through three normally closed motor-operated valves into the feedwater system through the non-safety-related portion of the CUW tie-in lines to the feedwater system.

DCD Revision 3, Tier 2, Table 6.2-9, "Secondary Containment Penetration List," identifies two new penetrations related to AFI system in the secondary containment.

The AFI system and its power supplies are non-safety grade, except for portions of the I&C (as discussed in Section 7.7 of this SER). The power supply for the pump and motor-operated valves are non-safety-related, and are physically separated from the emergency power supplies.

The system can be operated from the AFI pump house. The applicant states that this will ensure that the injection can be initiated within 30 minutes after the loss of normal makeup systems to provide sufficient cooling. In addition, the operator is provided with capability to control flow from the AFI pump house by throttling a motor-operated valve located in the pump room.

In the initial DCD Amendment Table 1.9-1, "Summary of ABWR Standard Plant COL License Information," the applicant identified the following COL license information items:

| ITEM NO | SUBJECT |
|---------|---------|
| 9.39 | Power Supply for Alternate Feedwater Injection Equipment |
| 9.40 | Test and Surveillance Intervals for Alternate Feedwater Injection Equipment, |

In the DCD amendment dated May 12, 2010 (ML101340548) the applicant deleted the COL license information items listed above.

ITAAC: The discussion of ITAAC is in Section 19S.4.5 of this SER.

Technical Specifications: There are no technical specification (TS) requirements associated with the AFI System.

### 9.5.14.3 Regulatory Basis

The regulatory basis and acceptance criteria for reviewing the AFI system are summarized below:

1. GDC 2, "Design bases for protection against natural phenomena," as it relates to safety-related SSCs that must be protected from, or capable of withstanding, the effects of such natural phenomena as earthquakes, tornadoes, hurricanes, and floods as described in DCD Chapters 2 and 3.

2. GDC 4, "Environmental and dynamic effects design bases," as it relates to safety-related SSCs that must be protected from, or capable of withstanding, the effects of externally generated missiles, internally generated missiles, pipe whip and jet impingement forces associated with pipe breaks and dynamic effects associated with possible fluid flow instabilities (e.g., water hammers).

3. GDC 60, "Control of releases of radioactive materials to the environment," as it relates to the ability of the AFI system design to control releases of radioactive materials to the environment.

4. GDC 16 "Containment design," as it relates to reactor containment and associated systems establishing an essentially leak-tight barrier against the uncontrolled release of radioactivity to the environment.

### 9.5.14.4 Technical Evaluation

NRC staff reviewed the STPNOC's application to amend the ABWR Design Certification Rule of the ABWR DCD. The staff's review of the proposed COL License Information Items 9.39 and 9.40 in the initial DCD amendment determined that these items should be part of the ITAAC. The applicant's revised DCD amendment withdrew these COL license information items and included additional ITAAC. These changes are consistent with the staff's recommendation, and therefore acceptable.

The staff's acceptance of the AFI system is based on the compliance of the system design with the requirements of GDC 2, 4, 16, and 60.

The results of the staff's review are discussed below.

GDC 2, "Design bases for protection against natural phenomena"

The staff reviewed the AFI system for compliance with the requirements of GDC 2 with respect to the design for protection against the effect of natural phenomena such as earthquakes, tornados, hurricanes, and floods. Compliance with the requirements of GDC 2 is based on adherence to Position C.1 of RG 1.29, "Seismic Design Classification," for the safety-related portions of the system and Position C.2 for non-safety-related portions of the system.

The applicant states in Subsection 9.5.14.1 of the amendment that the AFI system discharge piping is routed underground or is otherwise protected from physical impact. Subsection 9.5.14.2 of the amendment states that the piping and components interfacing with the CUW system are the same quality as that of the system up to and including the second check valve.

The AFI system will be located in AFI pump house and its water source will be located at least 300 feet from the reactor building, control building and turbine building. The AFI system is non-safety-related, and the AFI pump house contains no safety-related SSCs. Therefore, any failure of the system housed in the AFI pump house will not result in the failure of any safety-related SSCs and will not adversely affect the safety-related function of the ABWR design.

The AFI water source, which is located outside of the AFI pump house, will contain a minimum of 1,136,000 liters (300,000 gallons) of water for the AFI system to use. There is no information included in the amendment on what impact the failure of this tank would have on the plant flooding analysis. Because the water source supplying the AFI system contains a large inventory and is non-seismic, the staff issued RAI 09.02.04-2 requesting the applicant to discuss the impact that a failure of the water source would have on safety-related equipment.

The applicant's response to RAI 09.02.04-2 dated February 25, 2010 (U7-C-STP-NRC-100040, ML100600410), addresses potential flooding from the AFI water source by stating that the water source for the AFI is required to be located at least 90 meters (300 feet) from the reactor building, and the separation will minimize the possibility that flooding associated with the failure of the 1,136,000 liters (300,000 gallon) water source will have on SSCs located in those buildings. The applicant adds that in ABWR DCD Subsection 3.4.1.1.1, the flood protection measures in the ABWR design guard against flooding from outside storage tanks that may rupture, which include the condensate storage tank (CST), which has a capacity twice that of the AFI water source. The applicant concludes that any flooding from the AFI water source is bounded by the existing flood analysis.

The staff reviewed the flood analysis in DCD Subsection 3.4.1.1 and agreed that external flooding from an AFI water source failure is bounded by the external flooding that could result from the possible failure of the CST that was analyzed in the flood analysis. The staff's concern in RAI 09.02.04-2 regarding the AFI water source failure is therefore resolved.

The AFI system discharges through three normally closed MOVs into the feedwater system through the non-safety-related portion of the CUW tie-in lines, which are located in the reactor building main steam tunnel. Because, the break of the AFI line in the steam tunnel could also have flooding implications, RAI 09.02.04-2 also requests the applicant to discuss how the AFI line break is addressed in the appropriate flood analysis.

The applicant's response to RAI 09.02.04-2 dated February 25, 2010 (ML100600410), states that an AFI line break in the main steam tunnel area is bounded by the feedwater high energy line break, which is analyzed in DCD Tier 2, Subsection 3.4.1.1.2. The applicant further indicates that a break in the main steam tunnel area will be contained in the Seismic Category I structure and will not flood any safety-related equipment in the reactor building and control building.

The staff reviewed the applicant's response and the flooding analysis information in DCD Tier 2, Subsection 3.4.1.1.2 and found that the response adequately addresses the staff's concern, because the bounding break in the main steam tunnel is the feedwater line break and if flooding occurs in the main steam tunnel, the water volume is kept inside the tunnel until operators are ready to pump it to the radwaste system. Therefore, no safety-related equipment will be affected. The applicant provides a markup of DCD Section 9.5.14.2 "Safety Evaluation" to incorporate the above changes into the DCD. The staff confirmed that the revised ABWR DCD AIA Amendment, Subsection 9.5.14.2 is revised as committed in the RAI response. The staff considers RAI 09.02-04-2 resolved and closed.

Based on the above review, the staff concluded that the AFI system meets the requirements of GDC 2 as it relates to protecting the system against the effects of natural phenomena.

GDC 4, "Environmental and dynamic effects design bases"

The staff reviewed the AFI system for compliance with the requirements of GDC 4, "Environmental and dynamic effects design bases," as they relate to the dynamic effects associated with possible fluid flow instabilities, including induced water hammer and the effects of pipe breaks. Compliance with the requirements of GDC 4 is based on identification of the essential portions of the system that are protected from dynamic effects, including internally and externally generated missiles, pipe whip and jet impingement due to high and moderate energy missiles, and meeting the guidance in Branch Technical Position (BTP) 10-2, "Design Guidelines for Avoiding Water Hammers in Steam Generators."

The AFI system is a non-safety-related system. The system is being reviewed against GDC 4 to ensure that safety-related SSCs are not affected due to the AFI system's operation or its proximity to safety-related SSC's. The applicant states in Subsection 9.5.14.1 of the DCD amendment application that the AFI system discharge piping is routed underground or is otherwise protected from physical impact. Subsection 9.5.14.2 of the amendment states that the piping and components that interface with the CUW system are the same quality as those of the system up to and including the second check valve.

The application does not address what impact if any the failure of an AFI line could have on safety-related equipment in the reactor building or main steam tunnel, or what design features or operational measures will be used to minimize or preclude water hammer events due to the operation of the AFI system. The staff requested the applicant to provide information to address the AFI line failure (RAI 09.02.04-4) and water hammer (RAI 09.02.04-3).

The applicant's response to RAI 09.02.04-4 dated February 25, 2010 (ML100600410), states that regarding the impact of a postulated AFI system break on safety-related systems, the AFI is only required for beyond-design-basis events and is powered off and unpressurized for all normal and design-basis events. The applicant also states that should a pipe break in the AFI system occur, the dynamic effects of the break on the plant's safety systems are bounded by the main steam and feedwater line breaks because the AFI lines are at much lower temperature and pressure and have smaller line sizes. The staff found the applicant's response acceptable, since it confirms that the system is normally isolated and unpressurized, and that other breaks (main steam and feedwater) are bounding in terms of the resulting dynamic effects. Therefore, RAI 09.02.04-4 is resolved and closed.

The applicant's response to RAI 09.02.04-3 dated February 25, 2010 (ML100600410), states that the AFI design includes a "keep–full" line to maintain the system full of water to preclude water hammer upon system initiation. The applicant also states that existing vent valves in the feedwater pipe can be used to fill and vent the AFI system to ensure that the piping is maintained full of water. The applicant states the AFI system will be added to the ABWR DCD Section 13.5 list of systems for which operating procedures will need to be written. The procedures will include provisions to throttle the flow at the start of the AFI pump to minimize the potential for water hammer.

In this response, the applicant also provides a markup of DCD Subsection 9.5.14.1, "System Description," and DCD Subsection 13.5.3.4, "Procedures included in Scope of Plan," to incorporate these changes into the revised DCD. The staff found the applicant's response to RAI 09.02.04-3 acceptable, because the applicant adequately addresses water hammer concerns with system design features ("keep-full" line) and through the use of procedural precautions used in the industry to reduce or minimize water hammer. The staff confirmed that the revised ABWR DCD AIA Amendment, Subsections 9.5.14.1 and 13.5.3.4 are revised as committed in the RAI response. The staff considers RAI 09.02.04-3 resolved and closed.

Based on the above review, the staff concluded that the AFI system meets the requirements of GDC 4 as they relate to the dynamic effects associated with possible fluid instabilities, including induced water hammer and the effects of pipe breaks.

GDC 60, "Control of releases of radioactive materials to the environment,"

NRC staff reviewed the design of the AFI system for compliance with the requirements of GDC 60 with respect to control of releases of radioactive materials. GDC 60 requires that a means be provided to control the release of radioactive materials in liquid effluents.

Information on how the control of release of radioactive materials to environment is accomplished for the AFI system was not included in the amendment. Therefore, the staff issued RAI 09.02.04-1 requesting the applicant to address how the AFI system complies with GDC 60.

The applicant's response to RAI 09.02.04-1 dated April 19, 2010 (U7-C-STP-NRC-100085, ML101120085), states that the AFI contains two check valves located in the reactor building main steam tunnel, and three normally-closed MOVs. The applicant adds that in the event that fluid from the CUW system should leak past the two check valves and the first MOV, a leakoff line is included in the AFI design that directs any leakage back to the reactor building's low-conductivity sump, and the existing leak detection and radiation monitoring can be used to monitor the leakage. The applicant provides a markup of DCD Section 9.5.14.2 "Safety Evaluation" to incorporate the above changes into the DCD. The staff confirmed that the revised ABWR DCD AIA Amendment, Subsection 9.5.14.2 is revised as committed in the RAI response. The staff considers RAI 09.02.04-1 resolved and closed.

The staff found the information provided by the applicant in the response to RAI 09.02.04-1 acceptable, because the applicant design provides barriers against radiation leakage from radioactive systems into the non radioactive AFI system. The design also provides for routing of AFI system leakage to the reactor building low conductivity sumps, and existing leak detection and radiation monitoring can be used to monitor the leakage.

Based on the information in the applicant's response to RAI 09.02.04-1 and the staff's review of Figure 9.5-6 of the amendment, the staff found that the AFI design meets the requirements of GDC 60, as they relate to the AFI having design provisions to control the release of radioactive materials into the environment.

GDC 16 "Containment design"

In addition, the staff reviewed the design for impact on the integrity of the reactor building which forms the secondary containment under GDC 16 as it relates to establishing an essentially leak-tight barrier against the uncontrolled release of radioactivity to the environment. The additional penetrations identified in Table 6.2-9 are fluid piping penetrations that are designed under containment isolation provisions. This change does not alter the staff conclusion as documented in NUREG-1503. The staff found the addition of two penetrations acceptable.

Technical Specifications: The DCD amendment application does not include a TS section for the AFI system. The staff reviewed the AFI system against 10CFR 50.36, "Technical specifications," and agreed that no additional TS are required for this system. Therefore, the staff found this aspect of the DCD acceptable.

### 9.5.14.5 Conclusion

NRC staff reviewed the AFI system as part of the STPNOC's application to amend the Design Certification Rule for the ABWR. The staff's review of the information in the application found that it provides adequate information supporting the AFI system functional design. The staff found the amended design acceptable because it meets the appropriate regulatory requirements, including GDC 2, protection from natural phenomena; GDC 4, protection against missiles and effects of pipe break; and GDC 60, control of releases of radioactive materials into the environment. NRC staff also reviewed the two new penetrations identified in Table 6.2-9. The staff found the change acceptable because it meets the regulatory requirements of GDC 16. The staff determined that there would be no adverse impacts from complying with the requirements for consideration of aircraft impacts on conclusions reached by the NRC in its review of the original ABWR design certification (NUREG-1503).

# 13  CONDUCT OF OPERATIONS

### 13.5.3.4       Procedures Included in Scope of Plant

### 13.5.3.4.1       Introduction

As part of the application, STPNOC proposes to include the AFI system within the scope of the plant operating procedures development plan so that appropriate procedures will be developed for the AFI system.  This section of the SER discusses the staff's evaluation of the proposed changes and the technical basis for the staff's acceptance of these proposed changes.

### 13.5.3.4.2       Summary of Application

In the initial submittal dated June 30, 2009 (ML092040048), the applicant proposed to include the systems described in Section A.3 of ANSI/ANS-3.2, within the scope of the plant operating procedures development plan.

To address the aircraft impact assessment required by 10 CFR 50.150, the applicant, in a letter dated June 17, 2010 (U7-C-STP-NRC-100139, ML101720306), proposes to include the AFI system within the scope of the plant operating procedures development plan.

### 13.5.3.4.3       Regulatory Basis

The regulatory basis is established in 10 CFR Part 52, and Appendix B to 10 CFR Part 50; the guidance of RG's 1.8, 1.33, 1.70, and 1.206; and the guidance of Section 13.5.2.1 of NUREG-0800, Revision 2, and ANSI/ANS-3.2.

### 13.5.3.4.4       Technical Evaluation

NRC staff reviewed the information related to the AIA application in DCD Revision 3. Section A.3 of ANSI/ANS-3.2 identifies systems to which the guidance of ANSI/ANS-3.2 is applicable.  The applicant is adding one system in response to the aircraft impact assessment, which is required by 10 CFR 50.150. This new system is properly identified in accordance with ANSI/ANS 3.2, and Section A.3 is the appropriate place to include the new system.  Also, this is acceptable to comply with 10 CFR 50.120, "Training and qualification of nuclear power plant personnel."

### 13.5.3.4.5       Conclusion

NRC staff reviewed the inclusion of the AFI system within the scope of the plant operating procedures development plan.  The staff's review found the applicant has adequately addressed the applicable requirements.  In addition, the staff determined that there would be no adverse impacts from complying with the requirements for consideration of aircraft impacts on conclusions reached by the NRC in its review of the original ABWR design certification (NUREG-1503).

# 14 INITIAL TEST PROGRAM

## 14.2.1 Initial Test Program (ITP)

### 14.2.1.1 Introduction

As part of the application, STPNOC proposes changes the Initial Test Program (ITP) preoperational tests. This section of the ABWR DCD addresses the series of tests categorized as construction, preoperational, or initial startup tests. The application to amend the Design Certification Rule for the ABWR changes the ITP preoperational tests to meet the requirements in 10 CFR 52.79(a)(28), and 10 CFR 50.150 as they relate to STPNOC's request to amend the ABWR standard design certification to comply with the AIA rule. This section of the safety evaluation discusses the staff's evaluation of the ITP of the proposed additions and the technical basis for the staff's acceptance of these proposed changes.

### 14.2.1.2 Summary of Application

NRC staff identified the following changes in DCD Section 14.2, "Initial Plant Test Program," to address testing requirements for AFI equipment.

Alternate Feedwater Injection System Preoperational Test

In Subsection 14.2.12.1.78, "AFI Preoperational Test," the ABWR DCD AIA Amendment application incorporates the "Purpose, Prerequisites, General Test Methods and Acceptance Criteria" for this test.

The purpose of the AFI system preoperational test is to include testing of equipment, pumps, valves, and I&C. The prerequisites for conducting this preoperational test include completion of construction tests, The Startup Control Group (SCG) review and approval of the test procedure and of the test initiation. This includes the availability of an AFI pump suction water source to the reactor vessel through feed water lines A and B to receive AFI injection flow, and verification of the availability of an appropriate electrical power source to support this test.

The General Test Methods and Acceptance Criteria include performance of tests on individual components in the AFI system. The tests are designated to demonstrate that the AFI system operates properly through appropriate AFI system design specifications using the following 12 test methods and acceptance criteria:

- Proper operation of I&C in the AFI system.

- Verification of various component alarms.

- Proper operation of all MOVs, including opening and closing the MOVs with the operating switch, MOV status indication, and travel timing, if applicable.

- Proper operation of AFI pumps and motors during continuous run tests.

- Acceptable pump net positive suction head (NPSH) under the most limiting design flow conditions.

- Verification that the AFI System can be operated normally at each mode and will satisfy the NPSH requirement by combining all components, piping, and instruments constituting this system.

- Proper AFI pump motor start sequence and actuation of protective devices.

- Proper operation of interlocks including operation of all components subject to interlocking.

- Proper operation of permissive, prohibit, and bypass functions.

- Proper system operation while powered from primary and alternate sources, including transfers, and in degraded modes for which the system is expected to remain operational.

- Acceptable pump/motor vibration levels and system piping movements during both transient and steady-state operations. This test can be performed in conjunction with the expansion, vibration, and dynamic effects preoperational test (Subsection 14.2.12.1.51).

- Proper operation of the pump discharge line keep-fill system and its ability to prevent damaging water hammer during system transients.

### 14.2.1.3 Regulatory Basis

The regulatory basis and acceptance criteria for reviewing ITP are in Section 14.2 of NUREG-0800. In particular, the applicable regulatory requirements and guidance for the ITP portion of the ABWR DCD Amendment Include 10 CFR 52.79(a)(28), as it relates to preoperational testing and initial operations; and the guidance in RG 1.68, "Initial Test Programs For Light Water-Cooled Nuclear Power Plant," dated March 2007 and RG 1.206, "Combined License Applications for Nuclear Power Plants," Section C.I.14, "Verification Programs," dated June 2007.

### 14.2.1.4 Technical Evaluation

In accordance with 10 CFR 52.79(a)(28), NUREG–0800 Section14.2, and the regulatory positions in RG 1.68 and RG 1.206 Section C.I.14, the NRC staff reviewed the DCD amendment for these proposed changes to preoperational tests in Tier 2, DCD Section 14.2, to verify that they will satisfy the ITP.

Based on the regulations in 10 CFR 52.79(a)(28), the guidance in NUREG–0800. Section 14.2, and the regulatory positions in RG 1.68 and RG 1.206, the staff evaluated the AFI preoperational test acceptance criteria. The staff determined that, the purpose, prerequisites, and 12 general test methods and acceptance criteria in Tier 2, DCD Subsection 14.2.12.1.78 met 10 CFR 52.79(a)(28), RG 1.68 and RG 1.206, Section C.1.14 since they adequately test feedwater design features needed to mitigate the AIA beyond design basis event; therefore; they are acceptable.

### 14.2.1.5 Conclusion

NRC staff reviewed the ITP of the STPNOC's application to amend the Design Certification Rule for the ABWR. The staff's review concluded that the addition of Subsection 14.2.12.1.78, "AFI System Preoperational Test," met the requirements in 10 CFR 50.150, 10 CFR 52.79(a)(28), and conforms with the guidance in NUREG–0800 Section 14.2 and the regulatory positions in RG 1.68 and RG 1.206 Section C.I.14. Therefore, the test is acceptable.

## 14.3    Tier 1 Selection Criteria and Processes

### 14.3.1    Introduction

As part of the application, STPNOC proposes to add ITAAC for the AFI system and to specify the protection afforded by the reactor building in DCD Tier 1. The Tier 1 information in the ABWR DCD consists of an introductory section; design descriptions and corresponding ITAAC for the systems of the design, design material applicable to multiple systems of the design, interface requirements, and site parameters for the ABWR design. Furthermore, the purpose of the ITAAC, which are part of the Tier 1 information, is to verify that a facility that references the design certification has been constructed and will operate in accordance with the design certification and the applicable requirements.

### 14.3.2    Summary of Application

On May 12, 2010, STPNOC submitted a revised application to amend the Design Certification Rule for the U.S ABWR. In this submittal, the applicant proposes changes to Section 2.11.24, "Alternate Feedwater Injection System," and Section 2.15.10, "Reactor Building," of the certified DCD Tier 1 Information.

### 14.3.3    Regulatory Basis

The regulatory basis and acceptance criteria for reviewing the ITAAC information are in Section 14.3 of NUREG–0800. In particular, the applicable regulatory requirements and guidance for the ITAAC portion of the ABWR DCD Amendment Include: 10 CFR 50.150, and 10 CFR 52.47(b)(1), as they relate to ITAAC; and the guidance in RG 1.206, "Combined License Applications for Nuclear Power Plants," Section C.I.14, "Verification Programs," dated June 2007.

### 14.3.4    Technical Evaluation

In the applicant's initial DCD amendment, the staff found that the applicant did not propose an ITAAC for the AFI system. The staff issued RAI 14.2-01 requesting the applicant to provide an ITAAC for the AFI system. In response to this RAI, the applicant revised DCD Tier 1, Section 2.11.24 and Table 2.11.24 to meet the ITAAC requirements in 10 CFR 52.47(b)(1). These ITAAC are evaluated in Section 19S.4.5 of this SER.

In Tier 1, DCD Section 2.15.10, the applicant added a statement emphasizing that the specific areas of the reactor building are analyzed for the effects of postulated impact of an aircraft as required by 10 CFR 50.150. The staff reviewed the applicant's statement and concluded that 1) it was consistent with other Tier 1 information provided for the reactor building, 2) it was consistent with associated Tier 2 information, and 3) it adequately described the performance characteristics of the reactor building in relation to aircraft impacts. Therefore, the staff found the additional Tier 1 information acceptable.

## 14.3.5    Conclusion

The staff reviewed and concluded that the applicant proposed ITAAC for the AFI system are acceptable to meet the requirements in 10 CFR 50.150, and 10 CFR 52.47(b)(1) and conforms with the guidance in NUREG–0800, Section 14.3.  For additional details, see section 19S.4.5 of this SER.

The staff reviewed and concluded that the applicant's proposed change in the DCD Tier 1, Section 2.15.for the reactor building is an appropriate Tier 1 description and is acceptable.

# 17 QUALITY ASSURANCE

## 17.1.19  Quality Assurance During the Design Phase

### 17.1.19.1 Introduction

As part of the application, STPNOC proposes revising Section 17.0, "Introduction," of the ABWR DCD and adding Section 17.1.19, "Aircraft Impact Assessment," to Section 17.1, "Quality Assurance During the Design Phase." STPNOC also submitted an application on September 20, 2007, for a COL for two ABWRs designated as STP Units 3 and 4. The proposed Section 17.1.19 of the amended DCD references the STP Units 3 and 4 QA Program Description (QAPD). However, this SER does not apply to the STP Units 3 and 4 COL application.

This section of the safety evaluation discusses the staff's evaluation of the QA program during the design phase of the proposed additions and the technical basis for the staff's acceptance of these proposed changes. Also this section discusses the staff's determination of whether the changes adversely impact conclusions reached by the NRC in its review of the original ABWR design certification.

### 17.1.19.2 Summary of Application

The proposed Section 17.1.19, "Aircraft Impact Assessment," of the amended DCD Tier 2, Revision 3 addresses the establishment and implementation of a QA Program applicable to the proposed amendment to the ABWR design certification. In Section 17.1.19, the applicant references the STP Units 3 and 4 QAPD, Revision 4, dated June 7, 2010. The STP Units 3 and 4 QAPD, Revision 4, is included in the revised ABWR DCD AIA amendment dated June 17, 2010, (ML101720306).

### 17.1.19.3 Regulatory Basis

The regulatory basis for accepting the resolution to STP DCD Tier 2, Revision 3, Section 17.1.19 is satisfied based on the Commission's regulatory requirements related to QA Programs, which are set forth in 10 CFR 52.47(a)(19) and Appendix B, "Quality Assurance Criteria for Nuclear Power Plants and Fuel Reprocessing Plants," to 10 CFR Part 50.

10 CFR 52.54(a)(8) requires that the applicant implement the quality assurance program described or referenced in the safety analysis report.

10 CFR 52.47(a)(19) requires, in part, that a DC application contains a description of the QAP applied to the design of the SSCs of the facility. 10 CFR 52.47(a)(19) further requires that the description of the QAP for a nuclear power plant include a discussion of how the applicable requirements of Appendix B were satisfied.

10 CFR Part 50, Appendix B establishes QA requirements for the design, fabrication, construction, and testing of safety-related SSCs of the facility. The pertinent requirements of Appendix B apply to all activities affecting the safety-related functions of those SSCs and include designing, purchasing, fabricating, handling, shipping, storing, cleaning, erecting, installing, inspecting, testing, operating, maintaining, repairing, refueling, and modifying SSCs.

## 17.1.19.4 Technical Evaluation

RAI 01-2 and the applicant's response to this RAI are described in Section 1.4 of this SER. STPNOC's supplemental response to RAI 01-2 dated April 8, 2010 (U7-C-STP-NRC100080, ML101040345), describes the applicant's QA approach to the aircraft impact assessment and to the design changes resulting from that assessment. The applicant also proposes revising ABWR DCD Section 17.0, "Introduction," and adding Section 17.1.19, "Aircraft Impact Assessment," to Section 17.1, "Quality Assurance During the Design Phase." These proposed changes are incorporated into the ABWR DCD AIA amendment dated June 17, 2010, (ML101720306).

NRC staff reviewed (a) the proposed DCD Tier 2 Revisions provided in the supplemental response to RAI 01-2; (b) STPNOC's application to amend the Design Certification Rule for the U.S. ABWR; and (c) STP Units 3 and 4 QAPD, Revision 3. The staff's review is based on 10 CFR Part 50, Appendix B and applicable provisions of SRP Section 17.5. SRP Section 17.5 provides an outline of a QA Program acceptable to the staff for DC applicants. The staff developed SRP Section 17.5 using the American Society of Mechanical Engineers (ASME) Standard NQA-1-1994, "Quality Assurance Requirements for Nuclear Facility Applications," supplemented by additional regulatory and industry guidance for nuclear operating facilities. SRP Section 17.5 also addresses additional quality assurance requirements in 10 CFR Part 50, Appendix A, GDC 1 and in 10 CFR 50.34(f)(3)(ii) and (iii). GDC 1, "Quality standards and records," requires that a QA Program be established and implemented. 10 CFR 50.34(f)(3)(ii) and (iii) specify design and construction QA requirements which must be addressed in a QAPD.

The introduction of the QAPD, Section 1.1, "Scope/Applicability," states that, "Portions of the QAPD also apply to design quality assurance activities conducted in support of Design Certification Aircraft Impact Amendment (AIA). Assessments for AIA are beyond design basis and therefore the quality assurance requirements, of 10 CFR Part 50, Appendix B, do not apply. Controls in accordance with NEI 07-13 are in place to establish the validity of analyses and supporting calculations, and documentation files are maintained with a complete set of analyses. Changes to the facility design as a result of the assessments are subject to the full design controls described in this QAPD."

NRC staff conducted an inspection April 19 through 22, 2010, to provide the staff with reasonable assurance that the QA Program has been adequately implemented. The NRC staff's observations, conclusions, and Notice of Violation (NOV) are documented in Inspection Report Nos. 05200012/2010201 and 05200013/2010201 (ML101470298). One violation was identified for ineffective corrective action related to the inadequate documentation of personnel training and qualification. With the exception of the violation, the NRC inspection team concluded that the STPNOC QA policies and procedures complied with the applicable requirements of 10 CFR Part 21 and Appendix B to 10 CFR Part 50. The NRC inspection team further concluded that STPNOC personnel adequately implemented these policies and procedures. Therefore, the staff found that, STPNOC adequately implemented the QA Program described in its application as required by 10 CFR 52.54(a)(8).

### 17.1.19.4.1    Organization

The STP Units 3 and 4 QAPD follows the guidance in SRP Section 17.5, paragraph II.A, related to organization. The QAPD describes and defines the responsibility and authority for planning, establishing, and implementing an effective overall QA Program. The QAPD also describes an organizational structure; functional responsibilities; levels of authority; and interfaces for establishing, executing, and verifying implementation of the QAPD. The QAPD establishes independence between the organization responsible for verifying a function and the organization that performs the function. In addition, the QAPD allows the STP management to size the QA organization commensurate with its assigned duties and responsibilities.

In addition, in the STP Units 3 and 4 QAPD, the applicant commits to comply with the quality standards for QA organizations described in NQA-1-1994, "Basic Requirement 1," and in Supplement 1S-1.

### 17.1.19.4.2  Quality Assurance Program

The applicant's QAPD follows the guidance for the QA Program in SRP Section 17.5, paragraph II.B. The QAPD establishes measures that implement a QA Program to ensure that the design of a nuclear power plant is in accordance with governing regulations and license requirements. The QA Program comprises planned and systematic actions that are necessary to provide confidence that the SSCs will perform their intended safety functions, including certain non-safety-related SSCs and activities that are significant contributors to plant safety. The QA Program requires the maintenance of a list or system that identifies SSCs and activities to which the QAPD applies.

The QAPD provides measures that assess at least once each year or at least once during the life of the activity, whichever is shorter, the adequacy of the QAPD and ensure its effective implementation. In addition, consistent with SRP Section 17.5 paragraph II.B.8, the QAPD applies a grace period of 90 days to activities that must be performed on a periodic basis. The next due date for the performance of an activity that invokes the 90-day grace period remains unchanged. The next due date for an activity performed before the scheduled due date is moved forward, so the interval prescribed for the performance of the activity is not exceeded.

The QAPD also follows the guidance in SRP Section 17.5, paragraphs II.S and II.T, for training. The QAPD describes measures that establish and maintain formal indoctrination and training programs for personnel performing, verifying, or maintaining activities within the scope of the QAPD to ensure that they achieve and maintain a suitable level of proficiency. The QAPD also provides the minimum training requirements for managers responsible for implementing the QAPD, in addition to the minimum training requirements for those individuals responsible for planning, implementing, and maintaining the QAPD.

In the QAPD, the applicant commits to comply with the qualification and training standards described in NQA-1-1994, Basic Requirement 2; and Supplements 2S-1, 2S-2, 2S-3, and 2S-4, with the following clarifications and alternatives:

- NQA-1-1994, Supplement 2S-1 includes use of the guidance in Appendix 2A-1 the same as if it were part of the Supplement. NRC staff evaluated this clarification and determined that it is consistent with a previous approval of NEI 06-14, "Quality Assurance Program Description," Revision 7 (ML092650695) and is therefore acceptable.

- NQA-1-1994 Supplement 2S-3 states that the prospective lead auditors must have participated in a minimum of five audits in the previous 3 years. As an alternative to this requirement, the QAPD proposes to follow the guidance in SRP Section 17.5 paragraph II.S.4.c, which states that the prospective lead auditor shall demonstrate an ability to properly conduct the audit process (as implemented by the company), to effectively lead an audit team, and to effectively organize and report results, including participation in at least one nuclear audit within the year preceding the date of qualification. NRC staff evaluated this proposed alternative and determined that it is consistent with the regulation in 10 CFR Part 50, Appendix B, Criterion II. Therefore, the staff concluded that this alternative is acceptable.

### 17.1.19.4.3    Design Control

The applicant's QAPD follows the guidance of SRP Section 17.5, paragraph II.C for design control. The QAPD establishes the necessary measures that control the design and design changes that are subject to the QAPD provisions. The QAPD design process includes provisions for controlling design inputs, outputs, changes, interfaces, records, and organizational interfaces with the applicant and suppliers. These provisions ensure that the design inputs (i.e., design bases and the performance, regulatory, quality, and quality verification requirements) are correctly translated into design outputs (i.e., analyses, specifications, drawings, procedures, and instructions). In addition, the QAPD provides for individuals knowledgeable about QA principles to review design documents to ensure that they contain the necessary QA requirements.

In the QAPD, the applicant commits to comply with the quality standards described in NQA-1-1994, Basic Requirement 3 and Supplement 3S-1, to establish the program for design control and verification. The applicant also commits to comply with the requirements of Subpart 2.20 for the subsurface investigation requirements and Subpart 2.7 for the standards for computer software QA controls. NRC staff found these commitments acceptable.

### 17.1.19.4.4    Procurement Document Control

The applicant's QAPD follows the guidance in SRP Section 17.5 paragraph II.D, for procurement document control. The QAPD establishes the necessary administrative controls and processes to ensure that procurement documents include or reference applicable regulatory, technical, and QA Program requirements. As noted in SRP Section 17.5 paragraph II.D.1, the applicable technical, regulatory, administrative, quality, and reporting requirements (such as specifications, codes, standards, tests, inspections, special processes, and the regulation in 10 CFR Part 21, "Reporting of Defects and Noncompliance,") are invoked for the procurement of items and services.

In the QAPD, the applicant commits to comply with the quality standards described in NQA-1-1994, Basic Requirement 4 and Supplement 4S-1, with the following alternatives and commitment.

- NQA-1-1994 Supplement 4S-1, Section 2.3 states that procurement documents must require suppliers to have a documented QA Program that implements NQA-1-1994, Part I.

  - As an alternative to this requirement, the QAPD proposes that suppliers have a documented QA Program that meets the requirements of Appendix B to 10 CFR Part 50, as applicable to the circumstances of the procurement. NRC staff evaluated this proposed alternative and determined that it is consistent with Appendix B, Criterion IV, "Procurement Document Control." Therefore, the staff concluded that this alternative is acceptable.

  - As an alternative to this requirement, the QAPD proposes that procurement documents allow suppliers to work under the applicant's QAPD, including implementing procedures, if suppliers do not have their own QA Program. NRC staff evaluated this proposed alternative and determined that the applicant's QAPD follows the guidance in SRP Section 17.5 paragraph II.G, "Control of Purchased Material, Equipment, and Services." Specifically, the QAPD provides measures to evaluate prospective suppliers so that only qualified suppliers are selected, acceptance actions are performed for procured products and services, and suppliers are periodically audited and evaluated to ensure that qualified suppliers continue to provide acceptable products and services. Therefore, the staff concluded that this alternative is acceptable.

- NQA-1-1994 Supplement 4S-1, Section 3 states that procurement documents are to be reviewed before awarding the contract. As an alternative to this requirement, the QAPD proposes to conduct the QA review of procurement documents through the review of the applicable procurement specifications, including the technical and quality procurement requirements, before awarding the contract. In addition, procurement document changes (e.g., scope, technical, or quality requirements) will also receive a QA review. NRC staff evaluated this proposed alternative and determined that it provides an adequate QA review of procurement documents before awarding the contract and after any change. Therefore, the staff concluded that this alternative is acceptable.

- In the QAPD, the applicant commits to ensure that procurement documents prepared for commercial-grade items and procured for use as safety-related items shall contain technical and quality requirements, so that the procured item can be appropriately dedicated. NRC staff evaluated this proposed commitment and determined that it is consistent with staff guidance in Generic Letter 89-02, "Actions to Improve the Detection of Counterfeit and Fraudulently Marked Products," dated March 21, 1989 (Reference 3); and Generic Letter 91-05, "Licensee Commercial-Grade Procurement and Dedication Programs," dated April 9, 1991 (Reference 4); as delineated in SRP Section 17.5, paragraphs II.U.1.d and II.U.1.e. Therefore, the staff concluded that this commitment is acceptable.

### 17.1.19.4.5    Instructions, Procedures, and Drawings

The applicant's QAPD follows the guidance in SRP Section 17.5 paragraph II.E for instructions, procedures, and drawings. The QAPD establishes the necessary measures and governing procedures to ensure that activities affecting quality are prescribed by and performed in accordance with documented instructions, procedures, and drawings.

In the QAPD, the applicant commits to comply with the quality standards for instructions, procedures, and drawings described in NQA-1-1994, Basic Requirement 5 to establish procedural controls.

### 17.1.19.4.6    Document Control

The applicant's QAPD follows the guidance of SRP Section 17.5, paragraph II.F for document control. The QAPD establishes the necessary measures and governing procedures to control the preparation, review, approval, issuance, and revision of documents that specify quality requirements or prescribe measures for controlling activities affecting quality, including organizational interfaces. The QAPD provides measures to ensure that the same organization that performed the original review and approval should also review and approve revisions or changes to documents, unless other organizations are specifically designated. A listing of all controlled documents identifying the currently approved revision or date is maintained so personnel can readily determine the appropriate document for use.

In the QAPD, the applicant commits to comply with the quality standards for document control described in NQA-1-1994, Basic Requirement 6 and Supplement 6S-1, to establish provisions for document control.

### 17.1.19.4.7 Control of Purchased Material, Equipment, and Services

The applicant's QAPD follows the guidance in SRP Section 17.5, paragraph II.G for the control of purchased material, equipment, and services. The QAPD establishes the necessary measures and governing procedures to control the procurement of items and services that ensure conformance to specified requirements. The program provides measures to evaluate prospective suppliers so that only qualified suppliers are selected. In addition, the program requires the suppliers to be periodically audited and evaluated to ensure that qualified suppliers continue to provide acceptable products and services.

The program provides acceptance actions such as source verification, receipt inspection, pre-and post-installation tests, and the review of documentation such as certificates of conformance to ensure that procurement, inspection, and testing requirements have been satisfied before relying on the item to perform its intended safety function. Purchased items and services are subject to quality and technical requirements at least equivalent to those specified for original equipment—or by properly reviewed and approved revisions—to ensure that the items are suitable for the intended service and are of acceptable quality, consistent with their effects on safety.

In the QAPD, the applicant commits to comply with the quality standards for the control of purchased material, equipment, and services described in NQA-1-1994, Basic Requirement 7 and Supplement 7S-1, to establish procurement verification control with the following exceptions and alternatives.

- NQA-1-1994, Basic Requirement 7 and Supplement 7S-1, states that procurement sources and the performance of suppliers are to be evaluated. As an exception to these requirements, the QAPD proposes that other 10 CFR Part 50 licensees (other than the STP), authorized nuclear inspection agencies, the National Institute of Standards and Technology (NIST), and other State and Federal agencies that may provide items or services to STP are not required to be evaluated or audited.

  NRC staff acknowledged that 10 CFR Part 50 licensees, authorized nuclear inspection agencies, the National Voluntary Laboratory Accreditation Program (NVLAP) administered by NIST, and other State and Federal agencies perform work under quality programs acceptable to the NRC and no additional audits or evaluations are required. However, STPNOC remains responsible for ensuring that procured items or services conform to the Appendix B program, the applicable ASME Boiler and Pressure Vessel Code requirements, and other regulatory requirements and commitments. The applicant also remains responsible for ensuring that items or services are suitable for the intended application and for documenting the evaluation that supports this conclusion. The staff determined that this proposed exception provides an appropriate level of quality and safety and is therefore acceptable, as documented in a previous SER (ML003693241).

- SRP Section 17.5, paragraph II.L.8 establishes provisions for the procurement of commercial-grade calibration services for safety-related applications. As an exception to these provisions, the QAPD proposes not to require procurement source evaluation and selection measures, provided that all of the following conditions are met:

- Purchase documents will impose additional technical and administrative requirements to satisfy QAPD and technical requirements.

- Purchase documents will require the reporting of as-found calibration data, when calibrated items are found to be out of tolerance.

- A documented review of the supplier's accreditation will be performed and will include a verification of the following:

  (1) The calibration laboratory holds a domestic accreditation by any one of the following accrediting bodies, which are recognized by the International Laboratory Accreditation Cooperation (ILAC) Mutual Recognition Arrangement (MRA):

      a. NVLAP, administered by NIST
      b. American Association for Laboratory Accreditation (A2LA)
      c. ACLASS Accreditation Services (ACLASS)
      d. International Accreditation Service (IAS)
      e. Laboratory Accreditation Bureau (L-A-B)

  (2) The accreditation encompasses American Nuclear Society/International Standardization Organization/International Electrotechnical Commission 17025, "General Requirements for the Competence of Testing and Calibration Laboratories."

  (3) The published scope of accreditation for the calibration laboratory covers the necessary measurement parameters, ranges, and uncertainties.

NRC staff evaluated the NVLAP and A2LA accreditation programs and found them both acceptable (ML052710224). The staff subsequently determined that the accreditation of the ACLASS, L-A-B, and IAS Programs is also recognized by the ILAC MRA and is therefore acceptable (ML073440472, ML081140564, and ML081330253).

- NQA-1-1994 Supplement 7S-1, Section 8.1 states that documentary evidence that items conform to procurement documents shall be available at the nuclear facility site before installation or use. As an alternative to the requirement for procurement documentary evidence to be available at the nuclear facility site during construction, the QAPD proposes that documentary evidence may be stored in physical form or in electronic media, under the control of STP or its supplier, at a location other than the nuclear facility site, as long as the documents can be accessed at the nuclear facility site during construction. After the construction is complete, sufficient documentary evidence will be available to the licensee to support operations.

The staff determined that implementation of this alternative would allow access to and review of the necessary procurement documentary evidence at the nuclear facility site, both before installation and after use. Therefore, the staff concluded that this alternative is acceptable.

- As an alternative to the requirements for the control of commercial-grade items and services in NQA-1-1994 Supplement 7S-1 Section 10, the applicant commits in the QAPD to follow NRC guidance discussed in Generic Letters 89-02 and 91-05. In SRP Section 17.5, paragraphs II.U.1.d and II.U.1.e provide guidance to establish and describe special quality verification requirements in applicable documents to ensure that the commercially procured items will perform satisfactorily in service. In addition, the documents should provide for determining critical characteristics, technical evaluations, receipt requirements, and quality evaluations of the items to ensure that the items are suitable for their intended use.

The staff determined that this alternative will improve the likelihood of detecting counterfeit and fraudulently marked products and will improve the commercial-grade dedication programs. Therefore, the staff concluded that this alternative is acceptable.

### 17.1.19.4.8    Identification and Control of Materials, Parts, and Components

This element is not applicable to the DC amendment application and has not been reviewed or approved by the NRC staff.

### 17.1.19.4.9    Control of Special Processes

This element is not applicable to the DC amendment application and has not been reviewed or approved by the NRC staff.

### 17.1.19.4.10    Inspection

The applicant's QAPD follows the guidance in SRP Section 17.5, paragraph II.J for inspections. The QAPD establishes the necessary measures to implement inspections to ensure that items, services, and activities affecting safety meet established requirements and conform to applicable documented specifications, instructions, procedures, and design documents. The inspection program establishes requirements for planning inspections, determining applicable acceptance criteria, setting the frequency of inspections, and identifying special tools needed to perform the inspection. Properly qualified personnel independent of those who perform or directly supervise the work are required to perform the inspections.

In the QAPD, the applicant commits to comply with the quality standards for inspections described in NQA-1-1994, Basic Requirement 10, Supplement 10S-1, and Subparts 2.4, 2.5, and 2.8 to establish inspection requirements with the following commitment and alternative:

- NQA-1-1994, Subpart 2.4 requires the use of the IEEE Std 336-1985, "IEEE Standard Installation, Inspection, and Testing Requirements for Power, Instrumentation, and Control Equipment at Nuclear Facilities." IEEE Std 336-1985 refers to IEEE Std 498-1985, "IEEE Standard Requirements for the Calibration and Control of Measuring and Test Equipment Used in Nuclear Facilities." Each of these standards uses the definition of safety systems equipment from IEEE Std 603-1980, "IEEE Standard Criteria for Safety Systems for Nuclear Power Generating Stations." IEEE Std 603-1980 defines "safety system" as

    Those systems (the reactor trip system, an engineered safety feature, or both, including all their auxiliary supporting features and other auxiliary feature) which provide a safety function. A safety system is comprised of more than one safety group of which any one safety group can provide the safety function.

In the QAPD, the applicant must commit to the definition of safety systems equipment from IEEE Std 603-1980 to appropriately implement NQA-1-1994, Subpart 2.4. In the QAPD, the applicant commits to the definition of safety systems equipment from IEEE Std 603-1980 but does not commit to the balance of IEEE Std 603-1980. This definition applies only to equipment in the context of NQA-1-1994, Subpart 2.4. NRC staff determined that the use of the definition of safety systems equipment is acceptable because it is consistent with the requirements of NQA-1-1994, Subpart 2.4.

### 17.1.19.4.11    Test Control

The applicant's QAPD follows the guidance in SRP Section 17.5, paragraph II.K for test control. The QAPD establishes the necessary measures and governing provisions to demonstrate that items subject to the provisions of the QAPD will perform satisfactorily in service; that the plant can be operated safely as designed; and that the operation of the plant, as a whole, is satisfactory.

In the QAPD, the applicant commits to comply with the quality standards for test control described in NQA-1-1994, Basic Requirement 11, and Supplement 11S-1 to establish provisions for testing.

Furthermore, in the QAPD, the applicant commits to comply with the quality standards for software test control described in NQA-1-1994, Supplement 11S-2 and Subpart 2.7 to establish provisions ensuring that computer software used in applications affecting safety will be prepared, documented, verified, tested, and used so that the expected outputs are obtained and the configuration control is maintained.

### 17.1.19.4.12    Control of Measuring and Test Equipment

The applicant's QAPD follows the guidance in SRP Section 17.5, paragraph II.L for the control of measuring and test equipment (M&TE).  The QAPD establishes the necessary measures to control the calibration, maintenance, and use of M&TE that provide information important to a safe plant operation.

In the QAPD, the applicant commits to comply with the quality standards for M&TE described in NQA-1-1994, Basic Requirement 12, and Supplement 12S-1 to establish provisions for control of M&TE with the following clarification and exception:

- The QAPD clarifies that the out-of-calibration conditions described in paragraph 3.2 of Supplement 12S-1 of NQA-1-1994 refer to cases where the M&TE are found to be out of the required accuracy limits (i.e., out of tolerance) during calibration.  NRC staff determined that the clarification for the out-of-calibration conditions is consistent with Supplement 12S-1. Therefore, the staff concluded that this clarification is acceptable.

- As an alternative to NQA-1-1994 Subpart 2.4 Section 7.2.1, "Calibration Labeling Requirements," the QAPD proposes that when it is impossible or impractical to mark equipment with the required calibration information because of equipment size or configuration, the required calibration information will be documented and traceable to the equipment.  NRC staff determined that this alternative is consistent with the guidance in SRP 17.5, paragraph II.L.3.  Therefore, the staff concluded that this alternative is acceptable.

### 17.1.19.4.13    Handling, Storage, and Shipping

This element is not applicable to the DC amendment application and has not been reviewed or approved by the NRC staff.

### 17.1.19.4.14    Inspection, Test, and Operating Status

This element is not applicable to the DC amendment application and has not been reviewed or approved by the staff.

### 17.1.19.4.15    Nonconforming Materials, Parts, or Components

The applicant's QAPD follows the guidance in SRP Section 17.5, paragraph II.O for nonconforming materials, parts, or components.  The QAPD establishes the necessary measures to control items, including services, that do not conform to specified requirements to prevent inadvertent installation or use.  Nonconformances are evaluated for their impact on the operability of quality SSCs to ensure that the final condition does not adversely affect the safety, operation, or maintenance of the item or service.  The results of evaluations of conditions that adversely affect quality are analyzed to identify quality trends documented and reported to upper management, in accordance with the applicable procedures.

In addition, the QAPD establishes the necessary measures to implement the requirements of 10 CFR Part 52, 10 CFR 50.55, and 10 CFR Part 21, as applicable.

In the QAPD, the applicant commits to comply with the standards of quality for nonconforming materials, parts, or components described in NQA-1-1994, Basic Requirement 15, and Supplement 15S-1 to establish measures for nonconforming materials.

### 17.1.19.4.16    Corrective Action

The applicant's QAPD follows the guidance in SRP Section 17.5 paragraph II.P, for Corrective Action Programs.  The QAPD establishes the necessary measures to promptly identify, control, document, classify, and correct conditions that adversely affect quality.  The QAPD requires personnel to identify known conditions that adversely affect quality.  Reports of conditions that adversely affect quality are analyzed to identify trends.  Significant conditions that adversely affect quality are documented and reported to responsible management.  In the case of suppliers working on safety-related activities or similar situations, the applicant may delegate specific responsibilities for the Corrective Action Program, but the applicant maintains responsibility for the program's effectiveness.

In addition, the QAPD establishes the necessary measures to implement the requirements of 10 CFR Part 52, 10 CFR 50.55, and 10 CFR Part 21, as applicable.

In the QAPD, the applicant commits to comply with the standards of quality for corrective actions described in NQA-1-1994 Basic Requirement 16 to establish a Corrective Action Program.

### 17.1.19.4.17    Quality Assurance Records

The applicant's QAPD follows the guidance in SRP Section 17.5, paragraph II.Q for QA records.  The QAPD establishes the necessary measures to ensure that sufficient records of items and activities affecting quality are generated, identified, retained, maintained, and able to be retrieved.

In establishing measures to ensure that sufficient records of completed items and activities affecting quality are appropriately stored, the QAPD commits Regulatory Position C.2 and Table 1 of RG 1.28, Revision 3, for records and retention times.

Concerning the use of storage and retrieval systems for electronic records, the QAPD complies with the NRC guidance in Generic Letter 88-18, "Proposed Final NRC Generic Letter 88-18, Supplement 1," "Guidance on Managing Quality Assurance Records in Electronic Media," dated September 13, 1999; Regulatory Issue Summary 2000-18, "Guidance on Managing Quality Assurance Records in Electronic Media," dated October 23, 2000; and associated Nuclear Information and Records Management Association (NIRMA) guidelines TG 11-1998, TG 15-1998, TG 16-1998, and TG 21-1998.

In the QAPD, the applicant commits to comply with the standards for quality of QA records described in NQA-1-1994, Basic Requirement 17, and Supplement 17S-1 to establish provisions for records with the following alternative:

- In NQA-1-1994 Supplement 17S-1, Section 4.2(b) states that records must be firmly attached in binders or placed in folders or envelopes for storage in steel file cabinets or on shelving in containers. As an alternative to this requirement, the QAPD proposes that hard-copy records be stored in steel cabinets or on shelving in containers, except that methods other than binders, folders, or envelopes may be used to organize records for storage.

NRC staff determined that this alternative is acceptable, as documented in a previous SER (ML052360625).

### 17.1.19.4.18    Quality Assurance Audits

The applicant's QAPD follows the guidance of SRP Section 17.5, paragraph II.R for QA audits. The QAPD establishes the necessary measures to implement audits to verify that activities covered by the QAPD are performed in conformance with documented requirements. The Audit Program is reviewed for effectiveness as part of the overall audit process.

In the QAPD, the applicant provides for conducting periodic internal and external audits. Internal audits are conducted to determine the adequacy of program and procedures, and to determine if they are meaningful and comply with the overall QAP. Internal audits are performed with a frequency to ensure that an audit of all applicable QA program elements is completed within a period of once per calendar year or at least once during the life of the activity, whichever is shorter. External audits determine the adequacy of a supplier's or contractor's QA Program.

The applicant ensures that audits are documented and audit results are reviewed. In accordance with the QAPD, the applicant will respond to all audit findings and initiate appropriate corrective actions. In addition, where corrective actions are indicated, the applicant documents the follow-up of applicable areas through inspections, reviews, re-audits, or other appropriate means to verify the implementation of assigned corrective actions.

In the QAPD, the applicant commits to comply with the quality standards for QA audits described in NQA-1-1994, Basic Requirement 18, and Supplement 18S-1 to establish an independent Audit Program.

### 17.1.19.4.19    Non-Safety-Related SSC Quality Assurance Controls

### 17.1.19.4.19.1    Non-Safety-Related SSCs – Significant Contributors to Plant Safety

The applicant's QAPD follows the guidance of SRP Section 17.5, paragraph II.V.1 on controls related to nonsafety-related SSCs. The QAPD establishes program controls applied to nonsafety-related SSCs that are significant contributors to plant safety but are not covered by

Appendix B. The QAPD applies specific controls to these items in a selected manner, targeting the characteristics or critical attributes that render the SSC a significant contributor to plant safety consistent with applicable sections of the QAPD.

### 17.1.19.4.19.2   Non-Safety-Related SSCs Credited for Regulatory Events

The applicant's QAPD follows the guidance of SRP Section 17.5, paragraph II.V.2 to establish the quality requirements for non-safety-related SSCs credited for regulatory events.  In the QAPD, the applicant commits to comply with the following regulatory guidance:

- The applicant shall implement quality provisions for the fire protection system in accordance with Regulatory Position 1.7, "Quality Assurance," in RG 1.189, "Fire Protection for Operating Nuclear Power Plants," issued April 2001.

- The applicant shall implement quality provisions for anticipated transient without scram (ATWS) equipment in accordance with Generic Letter 85-06, "Quality Assurance Guidance for ATWS Equipment That Is Not Safety Related," issued January 1985.

- The applicant shall implement quality provisions for station blackout (SBO) equipment in accordance with Regulatory Position 3.5, "Quality Assurance and Specific Guidance for SBO Equipment That Is Not Safety Related," and Appendix A, "Quality Assurance Guidance for Non-Safety Systems and Equipment," in RG 1.155, "Station Blackout," issued August 1988.

### 17.1.19.4.20     Regulatory Commitments

The applicant's QAPD follows the guidance of SRP Section 17.5, paragraph II.U for describing regulatory commitments.  The QAPD establishes QA Program commitments.  In the QAPD, the applicant commits to comply with the following NRC regulatory guides and other QA standards to supplement and support the QAPD, with the noted clarifications and alternatives.

- RG 1.26, Revision 4, "Quality Group Classification and Standards for Water-, Steam, and Radioactive-Waste-Containing Components of Nuclear Power Plants," issued March 2007.

  The QAPD states that STPNOC conforms to the applicable regulatory positions through DCD Section 3.2.  Chapter 3 of this SER includes additional details.  NRC staff reviewed this clarification and found it consistent with the guidance in SRP 17.5 and therefore acceptable.

- RG 1.28 Revision 3, "Quality Assurance Program Requirements (Design and Construction)," issued August 1985.

  The QAPD identifies an alternative to Regulatory Position C.1 in Section 2.7 as accepted in a previous SER (ML070510300).  The applicant states that Regulatory Positions C.3.1 and C.3.2 are addressed in Sections 18.2 and 7.1 of the QAPD, respectively.  The staff reviewed these clarifications and alternatives and found them consistent with the guidance in SRP 17.5 and therefore acceptable.

  The QAPD states that Regulatory Position C.2 is addressed in Section 17.1

- RG 1.29 Revision 4, "Seismic Design Classification," issued March 2007.

  The QAPD states that STPNOC conforms to the applicable Regulatory Positions C.1 through C.3 through DCD Section 3.2.  Chapter 3 of this SER includes additional details.

The QAPD states that Regulatory Position C.4 is addressed by the QAPD requirements for safety-related activities. The staff reviewed these clarifications and found them consistent with the guidance in SRP 17.5 and therefore acceptable. Regulatory Position C.5 does not apply to AIA amendment activities.

- ASME NQA-1-1994, "Quality Assurance Requirements for Nuclear Facility Applications," Parts I and II, as described in foregoing sections.

- NIRMA technical guides, as described in Subsection 17.1.19.17 of the QAPD.

### 17.1.19.4.21 Additional Quality Assurance and Administrative Controls for the Plant Operational Phase

This element is not applicable to the DC amendment application and has not been reviewed or approved by the staff.

### 17.1.19.5 Conclusion

NRC staff reviewed the QAP of the STPNOC's application to amend the Design Certification Rule for the ABWR. The staff reviewed the proposed Section 17.1.19 of the application. Section 17.1.19 references the description of the QA Program applied to the DCD amendment activities associated with the STPNOC aircraft impact assessment using the STP Units 3 and 4 QAPD. The staff based the review of the STP Units 3 and 4 QAPD, Revision 4, on 10 CFR 52.47(a)(19); 10 CFR 52.54(a)(8); 10 CFR Part 50, Appendix B, "Quality Assurance Criteria for Nuclear Power Plants and Fuel Reprocessing Plants"; and SRP Section 17.5. The staff found that the QAPD used for the design phase of the proposed amendment to the ABWR Design Certification Rule is acceptable.

On the basis of the NRC staff's review of the STP Units 3 and 4 QAPD, the NRC staff concluded that:

- The STP Units 3 and 4 QAPD adequately describes the authority and responsibility of management and supervisory personnel, performance/verification personnel, and audit personnel.

- The STP Units 3 and 4 QAPD adequately provides for organizations and persons responsible for performing the verification and audit functions have the authority and independence to conduct their activities without undue influence from those directly responsible for costs and schedules.

- The STP Units 3 and 4 QAPD adequately applies to activities and items that are important to safety.

- The STP Units 3 and 4 QAPD adequately describes the program for the QA treatment of non-safety-related SSCs.

- The STP Units 3 and 4 QAPD adequately describes a philosophy and controls that, when properly implemented, comply with the requirements of 10 CFR 50.34(f)(3)(ii) and (iii) pursuant to 10 CFR 52.47(a)(8); Appendix B to 10 CFR Part 50 pursuant to 10 CFR 52.47(a)(19); and GDC 1.

Therefore, the NRC staff concluded that the STP Units 3 and 4 QAPD adequately describes the STPNOC's QA Program. Accordingly, the staff concluded that the STP Units 3 and 4 QAPD complies with the applicable NRC regulations and industry standards and can be used by STPNOC for DC amendment activities associated with the ABWR. The staff also concluded that the applicant adequately implemented the QA Program described in its application as required by 10 CFR 52.54(a)(8).

# 19S  DESIGN FEATURES FOR PROTECTION AGAINST A LARGE, COMMERCIAL AIRCRAFT IMPACT

## 19S.1    Introduction

As part of the application, STPNOC adds Appendix 19S to provide design features for protection against a large commercial aircraft impact.  This section of the SER describes the staff's evaluation of the description of design features and functional capabilities credited by the applicant to show that the facility can withstand the effects of a large, commercial aircraft impact.

The impact of a large commercial aircraft is a beyond-design-basis event.  Under 10 CFR 50.150, applicants for new nuclear power reactors[1] are required to perform an assessment of the effects on the designed facility of the impact of a large commercial aircraft. Applicants are required to submit a description of the design features and functional capabilities identified as a result of the assessment (key design features) in their DCD, together with a description of how the identified design features and functional capabilities show that the acceptance criteria in 10 CFR 50.150(a)(1) are met.  Applicants subject to 10 CFR 50.150 must make the complete aircraft impact assessment available for NRC inspection at the applicants' offices or their contractors' offices, upon an NRC request in accordance with 10 CFR 50.70, 10 CFR 50.71, and Section 161.c of the Atomic Energy Act of 1954, as amended.

## 19S.2    Summary of Application

In DCD Tier 2, Revision 3, Appendix 19S, the applicant states that an aircraft impact assessment in accordance with the requirements in 10 CFR 50.150(a)(1) was performed using the methodology described in NEI 07-13, "Methodology for Performing Aircraft Impact Assessments for New Plant Designs," Revision 7, as endorsed by the NRC in the draft regulatory guide, DG-1176.  Based on the results of the assessment, the applicant has identified a set of key design features to show that the acceptance criteria in 10 CFR 50.150(a)(1) are met.  In the July 12, 2010, AIA amendment submittal letter, the applicant states "all assumptions from the assessment documents, without which in our opinion the success criteria would not be met, have been included as key design features."  These key design features are reported in DCD Tier 2, Revision 3, Appendix 19S along with references to other sections of the DCD that provide additional details.  DCD Tier 2, Revision 3, Appendix 19S also contains descriptions of how the key design features show that the acceptance criteria in 10 CFR 50.150(a)(1) are met.

In the initial DCD Amendment Table 1.9-1, "Summary of ABWR Standard Plant COL License Information," the applicant identified the following COL license information items:

| ITEM No. | SUBJECT |
| --- | --- |
| 19.9k | Procedures for Use of Alternate Feedwater Injection |
| 19.9l | Procedures to Depressurize the Reactor Pressure Vessel (RPV) from the AFI Pump House |
| 19.9m | Verification of Environmental Conditions in the AFI Pump House |
| 19.9n | Description of Electrical Power Supply for AFI Equipment. |

---

[1]    "Applicants for new nuclear power reactors" is defined in the Statement of Considerations for the Aircraft Impact Rule (74 FR 28112, June 12, 2009).

In the DCD amendment dated May 12, 2010 (ML101340548), the applicant deleted the COL license information items listed above, since information was provided in the amended ABWR DCD as ITAAC.

## 19S.2.1   Description of Key Design Features

The credited key design features, their function(s), and references to sections containing the detailed descriptions are summarized below:

- The primary containment structure, as described in DCD Tier 2, Revision 3, Sections 3.8 and 3H.1, and Figure 1.2-1 protects the safety systems inside from impact.

- The location and design of the control building structure, as described in DCD Tier 2, Revision 3, Sections 1.2, 3.8.4, and 3H.2 and Figure 1.2-1 protects the north wall of the RB from impact.

- The location and design of the turbine building structure, as described in DCD Tier 1, Revision 3, Section 2.15.11 and Tier 2, Section 1.2 with Figures 1.2-1 and 1.2-24 through 1.2-31, protects the north wall of the control building and reactor building from impact.

- The location and design of the reactor building structure, as described in DCD Tier 2, Revision 3, Sections 1.2, 3.8.4, and 3H.1 and Figure 1.2-1 protects the south wall of the CB and primary containment from impact.

- The location and design of the spent fuel pool (SFP) and its supporting structure as described in DCD Tier 2, Revision 3, Sections 1.2 and 9.1 and Figure 1.2-12, protect the spent fuel pool from impact.

- The physical separation of the Class 1E emergency diesel generators and an independent power supply as described in DCD Tier 2, Revision 3, Section 9.5.14, prevent the loss of all electrical power to core cooling systems.

- The location and design of 3-hour fire barriers, including fire doors and watertight doors inside the reactor building and control building, as described in DCD Tier 2, Revision 3, Sections 9.5.1 and 9A.4 protect credited core cooling equipment from fire damage.

- The physical separation and design of the ECCS, as described in DCD Tier 2, Revision 3, Section 6.3 ensure core cooling.

- The design of the AFI system as described in DCD Tier 2, Revision 3, Section 9.5.14 ensures core cooling.

- The design of the containment overpressure protection system (COPS) as described in DCD Tier 2, Revision 3, Section 6.2.5 ensures core cooling.

## 19S.2.2   Description of How Regulatory Acceptance Criteria are Met

The acceptance criteria in 10 CFR 50.150(a)(1) are (1) the reactor core will remain cooled or the containment will remain intact, and (2) spent fuel pool cooling or spent fuel pool integrity is maintained.  The applicant states that it has met 10 CFR 50.150(a)(1) by maintaining both core cooling and spent fuel pool integrity.

As indicated in the amended ABWR DCD Tier 2, Revision 3, Appendix 19S, the applicant proposes to maintain core cooling using the safety-related and non-safety-related systems described in DCD Tier 2, Revision 3, Appendix 19S, which are specifically designed to ensure that the reactor can be shutdown and decay heat can be removed adequately from the reactor core. Some of this equipment is located (1) inside of the primary containment, (2) inside the reactor building, and (3) well away from the power block. Locations inside the primary containment are protected from structural, shock and fire damage by the design of the primary containment structure as well as the reactor building structure that limits the penetration of a large, commercial aircraft so that the primary containment is not perforated. Equipment inside the reactor building is protected by structural design features of the reactor building itself and by structures adjacent to the reactor building, including the turbine building and the control building. In addition, fire barriers are designed and located in the reactor building and control building to limit the spread of fire inside the buildings.

The applicant proposes to satisfy the spent fuel pool integrity acceptance criterion in 10 CFR 50.150(a)(1) due to the location and design of the spent fuel pool and its support structure. These key design features protect the structure from impact by a large commercial aircraft.

### 19S.2.3   Inspection, Test, Analysis, and Acceptance Criteria (ITAAC)

In accordance with 10 CFR 52.47(b)(1), the applicant has proposed the following ITAAC, as described in the amended ABWR DCD Tier 1, Revision 3, Section 2.11.24 and Table 2.11.24 (ML101340548) for key design features credited to meet 10 CFR 50.150:

ITAAC# 2.11.24-1– The basic configuration of the AFI system is as described in Section 2.11.24.

ITAAC# 2.11.24-2– The AFI pump is capable of injecting ≥800gpm into the RPV at the lowest SRV safety lift pressure

ITAAC# 2.11.24-3– The AFI system water supply has a minimum capacity of 300,000 gallons and is refillable.

ITAAC# 2.11.24-4– The AFI pump house is located a minimum of 300 feet from the nearest outside wall of each of the reactor building, control building and turbine building.

ITAAC# 2.11.24-5– The AFI water supply is located a minimum of 300 feet from the nearest outside wall of each of the reactor building, control building and turbine building.

ITAAC# 2.11.24-6– The AFI power supply is located a minimum of 300 feet from the nearest outside wall of each of the reactor building, control building and turbine building.

ITAAC# 2.11.24-7– Barriers exist, which qualify as the intervening structure as defined in NEI 07-13, Revision 7, between the AFI pump house and each of the reactor building, control building and turbine building.

ITAAC# 2.11.24-8– Barriers exist, which qualify as the intervening structure as defined in NEI 07-13, Revision 7, between the AFI water supply and each of the reactor building, control building and turbine building.

ITAAC# 2.11.24-9– Barriers exist, which qualify as the intervening structure as defined in NEI 07-13, Revision 7, between the AFI power supply and each of the reactor building, control building and turbine building.

ITAAC# 2.11.24-10– Instrumentation exists to provide information to the operator in the AFI pump house for reactor vessel water level, reactor pressure, suppression pool water level, and wetwell pressure.

ITAAC# 2.11.24-11– MOVs in the AFI system injection line operate as designed on a manual initiation signal.

## 19S.3    Regulatory Basis

NRC staff used the following relevant regulations and regulatory guidance to perform this review:

- 10 CFR 50.150(a)(1) requires that applicants perform a design specific assessment of the effects on the facility of the impact of a large, commercial aircraft.  Using realistic analyses, the applicant shall identify and incorporate into the design those design features and functional capabilities to show that, with reduced use of operator actions:  (i) The reactor core remains cooled, or the containment remains intact; and (ii) spent fuel cooling or spent fuel pool integrity is maintained.

- 10 CFR 50.150(b) requires that the final safety analysis report include a description of: (1) the design features and functional capabilities which the applicant has identified for inclusion in the design to show that the facility can withstand the effects of a large, commercial aircraft impact in accordance with 10 CFR 50.150(a)(1); and (2) how those design features and functional capabilities meet the assessment requirements of 10 CFR 50.150(a)(1).

- 10 CFR 52.47(b)(1) requires that a DC application contain the proposed ITAAC that are necessary and sufficient to provide reasonable assurance that, if the inspections, tests, and analyses are performed and the acceptance criteria met, a plant that incorporates the design certification has been constructed, and will operate in accordance with the design certification, the provisions of the Atomic Energy Act, and the NRC's regulations.

### 19S.3.1    Review Guidance

- DG-1176  "Guidance for the Assessment of Beyond-Design-Basis Aircraft Impacts," issued July 2009, provides guidance for meeting the requirements in 10 CFR 50.150(a), and specifically, documents NRC endorsement of the methodologies described in the industry guidance document, NEI 07-13, "Methodology  for Performing Aircraft Impact Assessments for New Plant Designs," Revision 7, issued May 2009.

- Statements of Consideration for the "Final Rule on Consideration of Aircraft Impacts for New Nuclear Power Reactors" (74 FR 28112, June 12, 2009) which indicates that for the NRC to conclude that the rule has been met, it must find that the applicant has performed an aircraft impact assessment reasonably formulated to identify design features and functional capabilities to show, with reduced use of operator action, that the acceptance criteria in 10 CFR 50.150(a)(1) are met.

Reasonably Formulated Assessment

The NRC considers an aircraft impact assessment performed by qualified personnel using a method that conforms to the guidance in NEI 07-13, Revision 7 to be a method that is reasonably formulated.  The NRC considers qualified personnel to be (1) an applicant who is the designer of the facility for which the aircraft impact assessment applies; or (2) an applicant's primary contractor for the aircraft impact assessment who has designed a nuclear power reactor facility either already licensed or certified by the NRC or currently under review by the NRC.

Cooling Criteria

The "reactor core cooling" criterion or "spent fuel pool cooling" criterion in 10 CFR 50.150(a)(1) is satisfied if design features have been included in the design of the plant to specifically perform that cooling function with reduced use of operator action.

Containment Criteria

The "intact containment" criterion in 10 CFR 50.150(a)(1) is satisfied if the containment: (1) will not be perforated by the impact of a large, commercial aircraft; and (2) will maintain ultimate pressure capability, given a core damage event until effective mitigation strategies can be implemented.  Effective mitigation strategies are those that provide, for an indefinite period of time, sufficient cooling to the damaged core or containment to limit temperature and pressure challenges below the ultimate pressure capability of the containment as defined in Section 19 of the DCD, Revision 3.

Spent Fuel Pool Integrity Criteria

The "spent fuel pool integrity" criterion in 10 CFR 50.150(a)(1) is satisfied if the impact of a large, commercial aircraft on the spent fuel pool wall or support structures would not result in leakage through the spent fuel pool liner below the required minimum water level of the pool.

Reduced Operator Action

The NRC considers the use of operator actions to be reduced when (1) all necessary actions to control the nuclear facility can be performed in the control room or at an alternate station containing equipment specifically designed for control purposes; and (2) a reduced amount of active operator intervention, if any, is required to meet the acceptance criteria in 10 CFR 50.150(a)(1).  Reductions in the use of operator action are measured relative to the actions required to address aircraft impact without the aircraft impact assessment rule in place (e.g., similar actions contained in operational programs in place at current operating reactor sites).

## 19S.4    Technical Evaluation

The staff's review of the proposed COL License Information Items 19.9k, 19.9l, 19.9m, and 19.9n in the initial DCD amendment determined that these items should be part of the ITAAC. The applicant's revised DCD amendment withdrew these COL license information items and included additional ITAAC. These changes are consistent with the staff's recommendation, and therefore acceptable.

NRC staff reviewed the applicant's description of key design features and the description of how the key design features show that the design meets the acceptance criteria in 10 CFR 50.150(a)(1). The staff's evaluation follows:

### 19S.4.1 Reasonably Formulated Assessment

The applicant states in the application that a design-specific assessment was performed with specific assumptions in NEI 07-13 Revision 7. The staff found this information incomplete because it was not clear that the applicant followed NEI 07-13 Revision 7 and NRC-endorsed guidance in whole or if there were any exceptions. The staff issued RAI 19-8 requesting the applicant to clarify whether the applicant fully followed NEI 07-13 or to state any exceptions. The applicant's response dated February 18, 2010 (U7-C-STP-NRC-100040, ML100550027), clarifies that the applicant followed NEI 07-13 with no exceptions. The analysis was performed by the applicant's subcontractor who has designed a nuclear power reactor facility either already licensed or certified by the NRC or currently under review by the NRC. Therefore, the staff found that the applicant has performed the assessment with a method that is reasonably formulated because the assessment was performed by a qualified organization, and it followed NEI 07-13 with no exceptions, and RAI 19-8 is therefore resolved. The staff verified that the applicant in the DCD AIA Amendment dated May 12, 2010 (U7-C-STP-NRC-100098, ML101340548) has followed NEI 07-13 without exception. Therefore, RAI 19-8 is closed.

### 19S.4.2 Key Design Features for Core Cooling

The key design features listed in the application that perform a core cooling-related function include (1) safety-related design features that are designed specifically to perform the core cooling function during normal power operation and following design-basis events initiated during power operation, and (2) non-safety-related design features that are designed specifically to support the core cooling function following the impact of a large commercial aircraft. The safety-related design features include the ECCS described in DCD Tier 2, Section 6.3; the SRVs described in DCD Tier 2, Section 5.2.2; the suppression pool described in DCD Tier 2, Section 6.2.1; and the COPS described DCD Tier 2, Section 6.2.5. The non-safety-related design features is the AFI system described in the application (Section 9.5.14). The preferred method of core cooling following the impact of a large commercial aircraft is to utilize normal safety-related injection systems and suppression pool cooling systems. However, if these systems are unavailable due to damage, the AFI system is used to add water to the reactor vessel and maintain the water level above the top of the active fuel, and heat is removed from the reactor vessel by relieving steam to the suppression pool via the SRVs. Steam is released to the outside atmosphere via the COPS after pressure in the suppression pool reaches the rupture pressure of the COPS rupture discs.

NRC staff considered the descriptions of the safety-related features as well as staff's review of the ability of these features to perform their design-basis safety functions documented in the FSER supporting certification of the ABWR design (NUREG–1503). During the review, the staff confirmed that all of these design features can be initiated and operated from the control room or at an alternate station containing equipment specifically designed for control purposes and require little, if any, additional operator intervention to maintain the core cooling function. Based on this information, the staff found that the applicant has adequately described the use of these systems to maintain core cooling following an impact from a large commercial aircraft when the reactor is shut down before the impact.

In the initial review of the applicant's descriptions, the staff noted that there was no description of design features or functional capabilities relied on to ensure that the acceptance criteria in 10 CFR 50.150(a)(1) are met, if the reactor fails to shut down following either an automatic or manual actuation of the reactor protection system. The staff issued RAI 19-3 requesting the applicant to either describe how those design features and/or functional capabilities that will be credited for core cooling are able to be successful if the reactor fails to shutdown, or identify design features that ensure that the reactor will be shut down following the impact of a large, commercial aircraft. The applicant's response to RAI 19-3 dated February 18, 2010 (U7-C-STP-NRC-100043, ML100550025), states that the design of the facility locates the hydraulic control units (HCU) for reactor scram outside of the damage footprint that could be caused by the impact of a large commercial aircraft; and, as a result, advance warning to successfully shut down the reactor manually before impact may be assumed in accordance with the guidance in NEI 07-13. NEI 07-13 also requires that an assessment be made to determine if the physical damage from an aircraft impact could prevent the reactor from being shutdown in cases where advanced warning is not available. In response to RAI 15.08-1 dated February 18, 2010 (ML100550027) the applicant states that ABWR has two divisions of HCUs and both are located below grade and are not subject to physical damage from an aircraft impact. The staff reviewed the description of the HCUs in DCD Tier 2, Revision 3 and found that the HCUs are located below ground level and outside of the structural damage footprint and are designed to fail in a safe manner, when subjected to fire (i.e., the loss of function results in reactor scram). The staff also reviewed NEI 07-13, Revision 7 and confirmed that the guidance indicates that an assumption of a successful scram may be made when the equipment required is not expected to be affected by the damage caused from an impact of a large commercial aircraft. The applicant's response indicates that the application will be modified to reflect this response. Therefore, based on the review described above, the staff found the applicant's response acceptable because it adequately describes how the design features will ensure the AIA criteria are met. The staff verified that the revised DCD AIA Amendment dated May 12, 2010 (ML101340548) states that the design of the facility identifies the hydraulic control units (HCU) for the reactor scram as outside of the damage footprint that could be caused by the impact of a large commercial aircraft. As a result, the operator would have advance warning to successfully shut down the reactor manually before impact may be assumed in accordance with the guidance in NEI 07-13. Therefore, RAI 19-3 is closed.

As discussed in the application, the AFI is designed to inject unheated water into the reactor vessel at high pressure (i.e., lowest opening pressure of any SRV) at a flow rate of 800 gpm. Considering the decay power of the ABWR, the staff found that the designated flow rate will be adequate to make up for a loss of inventory due to steam production from decay heat.

The applicant did not describe in the DCD the method for removing heat from the reactor vessel when the AFI system is being used to inject water. Consequently, the staff issued RAI 19-2 requesting the applicant to describe how heat is removed from the reactor vessel when the ECCS are unavailable and the AFI is being used to inject water into the reactor vessel. The applicant's response to RAI 19-2 dated February 18, 2010 (U7-C-STP-NRC-100043, ML100550025), states that SRV cycling at high pressure will relieve energy from the reactor vessel to the suppression pool, and as the suppression pool heats up, pressure in the containment will increase until the COPS rupture disc opens, which will relieve steam to the atmosphere. The staff found this response describes a viable heat removal method because it utilizes safety-related systems and components specifically designed to perform the functions needed to take steam from the reactor vessel and vent it to the outside atmosphere.

The applicant's response to RAI 19-2 also proposes a modification to the application that adds a more detailed description of the equipment used for decay heat removal and the heat removal pathways. The staff reviewed the proposed descriptions and confirmed that they completely describe the methods of heat removal and plant equipment relied upon to implement the methods. Therefore, the staff found the proposed modification acceptable. The staff verified that the applicant has included complete descriptions of the methods of heat removal and plant equipment relied upon to implement the methods in the revised application dated May 12, 2010 (ML101340548). The staff confirmed that the final DCD AIA Amendment provides the proposed DCD changes. Therefore, RAI 19-2 is resolved and closed.

The application states that the water source for the AFI will have a minimum capacity of 300,000 gallons of water. The staff issued RAI 19-7 requesting the applicant to provide the basis for choosing this capacity and if this capacity was large enough to support core cooling until mitigation measures can be implemented that provide long-term cooling. The applicant's response to RAI 19-7 dated February 18, 2010 (U7-C-STP-NRC-100043, ML100550025), states that the capacity was sized to provide sufficient make-up water to offset the calculated boil-off of water for 24 hours following scram from 100 percent power, assuming that the AFI is the only system available to provide core cooling. The applicant adds that 24 hours will be sufficient to replenish the AFI water source or initiate a firewater addition system for long-term cooling. The staff considers 24 hours to be a sufficient amount of time to implement mitigation measures for long-term core cooling for the following reason. Separate and apart from the requirements in 10 CFR 50.150, power reactor licensees are required, per 10 CFR 50.54(hh)(2), to develop and implement strategies for maintaining core cooling under conditions associated with the loss of large areas of the plant due to explosion or fire. The staff expects that licensee compliance with these requirements will result in implementation of mitigation measures within 24 hours. Therefore, the staff determined the applicant's technical basis for the capacity of the AFI water source acceptable. Therefore, RAI 19-7 is resolved and closed.

The staff issued RAI 19-5 requesting that the applicant to describe any support systems or components that the AFI system relies upon for successful operation, such as systems for pump cooling or lubrication. The purpose of this request is to ensure that the applicant has identified as key design features all pieces of equipment needed for core cooling. The applicant's response to RAI 19-5 dated February 18, 2010 (U7-C-STP-NRC-100043, ML100550025), states that the AFI system does not rely on any special or unique auxiliary support systems or equipment. The applicant explains that the AFI pump relies only on the water it is pumping for its own cooling and does not use auxiliary cooling equipment. The applicant adds that an oil pumping system is not used to lubricate the AFI pump. Lubrication is performed manually as part of the standard maintenance program for the pump. Based on the applicant's response, the staff found that with the exception of the electric power support system described in the application, there are no additional systems or components associated with the AFI system that need to be identified and described. The staff confirmed that the final DCD AIA Amendment provides the proposed DCD changes. Therefore, RAI 19-5 is resolved and closed.

Based on the description in the application, the AFI system includes dedicated I&Cs to allow operators to start and control the system and to monitor the pressure and water level in the reactor vessel to ensure that the system is performing properly. The staff found that the inclusion of these features in the design allows the AFI to be used to maintain core cooling with reduced operator action.

Based on the information in the application and the responses to the above RAIs, the staff found that the applicant has adequately described how core cooling will be maintained for at power scenarios using the AFI system in conjunction with the SRVs, the suppression pool, and the COPS.

In the initial review of the applicant's descriptions, the staff noted that the applicant did not include a description of the design features or functional capabilities relied upon to ensure that the acceptance criteria in 10 CFR 50.150(a)(1) are met, while the plant is shut down and the reactor core is being cooled via the shutdown cooling system with a large vent[2] in the primary system. The staff issued RAI 19-1 requesting the applicant to describe those design features and/or functional capabilities relied upon to ensure that the acceptance criteria in 10 CFR 50.150(a)(1) are met, while the plant is shut down and the reactor core is being cooled via the shutdown cooling system. The applicant's response to RAI 19-1 dated February 18, 2010 (U7-C-STP-NRC-100043, ML100550025), proposes a modification to the application stating that the existing RHR system will provide sufficient decay heat removal, but if the existing RHR system is unavailable as a result of the damage cause by the impact from a large commercial aircraft, decay heat removal will be accomplished by evaporation, with makeup provided by the AFI system. The staff considered the use of the RHR and AFI to be acceptable means of maintaining core cooling during shutdown conditions because those systems are designed to remove decay heat from the reactor and are capable of performing this function under plant conditions that exist with the a reactor shutdown for refueling. The staff verified that the revised DCD AIA Amendment dated May 12, 2010 (ML101340548) states that the existing RHR system will provide sufficient decay heat removal unless the existing RHR system is unavailable as a result of the damage cause by the impact from a large commercial aircraft, in which case, decay heat removal will be accomplished by evaporation with makeup provided by the AFI system. The staff confirmed that the final DCD AIA Amendment provides the proposed DCD changes. Therefore, RAI 19-1 is resolved and closed.

The applicant's description of the instrumentation included with the AFI system states that instrumentation exists to provide information to the operator in the AFI pump house for reactor vessel water level, reactor pressure, suppression pool water level, and wetwell pressure. The staff found this description acceptable because this is the minimum information needed to maintain core cooling by an operator during water injecting into the reactor vessel.

The applicant's description of the AFI system power supply states that it is designed to power the AFI pump and motor operated valves and will meet non-Class 1E design requirements for non-safety-related power supplies. The staff found this description acceptable because the pump and valves are the only components of the system requiring motive power and because key design features included to satisfy the requirements of 10 CFR 50.150 are not required to be safety related.

The staff considered the potential for re-criticality of the reactor core and thermal shock to the reactor vessel caused by cold water injected with the AFI system. The staff compared the description of the AFI system with the description of the RCIC system. The RCIC system is part of the certified ABWR design and designed to inject water as cold as 10°C from a storage tank into the reactor vessel when the reactor coolant system is at pressures above normal operating pressure. In addition, the Commission has determined that the certified ABWR design,

---

[2]    A large vent is one of sufficient size to ensure that gas, including steam, can be removed from the reactor coolant system and water can be added to it should a loss of decay heat removal or loss of inventory event occur when the system is in cold shutdown.

including the RCIC system, satisfies its requirements for emergency core cooling and fracture toughness. The staff found that the RCIC and the AFI systems are very similar in functional design. Therefore, it is reasonable to conclude that the AFI system can perform its core cooling function successfully without complications due to re-criticality or thermal shock.

During the initial review of the amendment the staff determined that the description of the AFI water source was not adequate. The staff issued RAI 19-4 requesting the applicant to provide additional descriptive information regarding the design of the water source, and request that this information be included in the amended DCD. In the RAI the staff stated that if the applicant intended to only specify the functional capability of the water source and leave it to the COL applicant referencing the DC to provide this information as part of its detailed design, then the applicant for design certification should propose an ITAAC specifying the acceptance criteria that must be met to ensure compliance with 10 CFR 50.150. Inclusion of the ITAAC would require the COL Holder to ensure that the design of the water source meets the acceptance criteria before operation. In the response to this RAI dated February 18, 2010 (ML100550025), the applicant indicates that it will provide an ITAAC for the AFI water source regarding the capacity of the water source and its location relative to structures in the power block. The applicant provides a proposed ITAAC for the AFI water source in the response to RAI 14.02-1 dated April 8, 2010 (ML101040254). The staff's review of the response to RAI 14.02-1, including the proposed ITAAC to be included in the DCD Tier 1, is documented in Section 19S.4.5 of this SER. The staff found the applicant's response to RAI 19-4 acceptable because the applicant has provided a proposed ITAAC that addresses the capacity of the water source and its location relative to structures in the power block. The staff reviewed the revised DCD AIA Amendment (ML101340548) and confirmed that the proposed changes are incorporated. Therefore, RAI 19-4 is resolved and closed.

The staff also determined that the description of the AFI power source was not adequate. The staff issued RAI 19-6 requesting the applicant to propose an ITAAC that addresses the location of the power source relative to structures in the power block and identify any time delay in providing power necessary to operate the AFI system. In the response to this RAI dated February 18, 2010 (ML100550025), the applicant indicated that it would provide an ITAAC for the AFI power source regarding the location of the power source relative to structures in the power block. The applicant provided a proposed ITAAC for the AFI power source in the response to RAI 14.02-1 (ML101040254). The staff's review of the response to RAI 14.02-1, including the proposed ITAAC to be included in the DCD Tier 1, is documented in Section 19S.4.5 of this SER. In the response, the applicant indicates that the time delay for providing power to the AFI was not addressed in the ITAAC because the design specification that injection with AFI be initiated within 30 minutes of an aircraft impact- described in Subsection 9.5.14.1 of the amendment- accounts for any delay in delivering motive power to the pump and valves in the AFI system. The staff found this response to be an acceptable rationale for not addressing the time delay for providing power to the AFI in the ITAAC because the time delay is captured in overall time delay which is described explicitly in the AFI design description which can be incorporated by reference in a COL application. In addition, the staff found the applicant's response to RAI 19-6 acceptable because the applicant has provided a proposed ITAAC that addresses the location of the power source relative to structures in the power block and an explanation for not addressing the power system time delay in the ITAAC which the staff finds acceptable, as discussed above. The staff reviewed the revised DCD AIA Amendment (ML101340548) and confirmed that the proposed changes are incorporated. Therefore, RAI 19-6 is resolved and closed.

Furthermore, the staff issued RAI 19-11 requesting the applicant to provide descriptive information and clarification on several statements in DCD Section 19S.4.2 regarding the AFI power source, the separation between the AFI power source, and the emergency power sources in the power block. In the response to this RAI dated February 25, 2010 (ML100600410), the applicant provides adequate clarification and indicates that it will provide an ITAAC for the AFI power source regarding the location of the power source relative to structures in the power block. The applicant provides a proposed ITAAC for the AFI power source in the response to RAI 14.02-1. The staff's review of the response to RAI 14.02-1, including the proposed ITAAC to be included in the DCD Tier 1, is documented in Section 19S.4.5 of this SER. The staff found the applicant's response to RAI 19-11 acceptable because the applicant has clarified statements in the amendment as requested by the NRC and provided a proposed ITAAC that addresses the location of the AFI power source relative to structures in the power block, including the normal emergency power sources. The staff reviewed the revised DCD AIA Amendment (ML101340548) and confirmed that the proposed changes are incorporated. Therefore, RAI 19-11 is resolved and closed.

### 19S.4.3 Key Design Features that Protect Core Cooling Design Features

*Fire Protection*

The key design features of the Fire Protection Program that protect the core cooling key design features include all of the 3-hour, fire-rated barriers and 5-psid barriers in the reactor building and the control building, as described in DCD Tier 2, Revision 3, Sections 9.5.1 and 9A.4, Appendix 19S and Figures 9A.4-4 through 9A.4-8. In the initial review of the descriptions provided by the applicant, the staff noted that the applicant did not include the overpressure ratings for each credited fire door. NRC staff issued RAI 19-12 requesting the applicant to identify, in the DCD, the overpressure ratings for each credited fire door. The applicant's response to RAI 19-12 in a letter dated February 18, 2010 (U7-C-STP-NRC-100040, ML100550027), identifies those doors that are identified as watertight doors, as shown in DCD Tier 2, Revision 3, Figure 1.2-8, which will be rated for at least 5 psid. In addition, the applicant clarifies that all other doors will not be rated at the minimum 5 psid. The staff considers watertight doors to have an equivalent rating of at least 5 psid, which provides an adequate means of confining deflagration overpressures as described in NEI 07-13 and an acceptable overpressure description. All watertight doors must maintain a fire rating that is consistent with the fire barrier where they reside.

In the initial review of the descriptions provided by the applicant, the staff noted that the applicant did not identify the reactor building R/B floor plugs as key design features. These floor plugs separate the ground floor from the lower elevations that contain ECCS equipment. The floor plugs were added to limit the fire spread and protect key core cooling features. The staff issued RAI 19-13 requesting the applicant to identify these floor plugs as key design features in the DCD. The applicant's response to RAI 19-13 in a letter dated February 18, 2010 (U7-C-STP-NRC-100043, ML100550025), states that it will include these floor plugs as key design features and to identify them as such in the DCD. The staff considered the applicant's responses acceptable. In the responses to both RAI 19-12 and RAI 19-13, the applicant provided a proposed DCD markup that the staff found acceptable. The staff confirmed that the revised DCD AIA Amendment provides the proposed DCD changes. Therefore, RAI 19-12 and 19-13 are resolved and closed.

During the review of DCD amendment submittal dated June 30, 2009 (ML092040048), NRC staff discovered an issue related to the NEI 07-13, Revision 7, "Fire Spread Rule" set.

Specifically, the applicant specified only the pressure rating of fire doors within the barriers that the applicant was crediting for withstanding overpressure effects. The staff's position is that the entire barrier needs to be rated at least 5-psid when taking credit for a single barrier. This includes fire doors, penetration seals, hatches, etc. The basis for the staff's position is that an overpressure may damage the components of the credited 3-hour fire-rated barrier, allowing fire to propagate beyond the credited barrier and beyond the fire damage footprint. In addition, proper identification and adequate description of these design features are required by 10 CFR 50.150. Therefore, the staff issued RAI 19-14 requesting the applicant to state within the DCD the pressure rating of all barriers and barrier components identified under the one barrier option of the NEI 07-13 Fire Spread Rule set. The applicant's response to RAI 19-14 dated June 21, 2010 (U7-C-STP-NRC-100152, ML101750070), states that Subsection 9.5.1.1.3 will be updated to state that the structural walls, floor, ceilings, and penetration seals and hatches within all fire barriers of the reactor building will be rated to withstand a 5 psid pressure differential. In the response to RAI 19-14, the applicant referenced a proposed DCD markup in a letter dated June 17, 2010 (U7-C-STP-NRC-100139, ML101720306) and provided a supplemental markup that the staff found acceptable. The proposed DCD includes the structural floor and ceilings as well as hatches in its discussion regarding the capability of certain fire protection features to withstand a 5 psid pressure differential. The staff confirmed that the final DCD AIA Amendment provides the proposed DCD changes. Therefore, RAI 19-14 is resolved and closed.

In the DCD amendment submittal dated June 30, 2009 (ML092040048), the applicant did not describe all fire doors adequately. The applicant described some doors as rated for 5-psid overpressure. These doors should also be rated as 3-hour fire-rated doors to meet the existing Fire Protection Program and fire hazards analysis requirements, as well as the guidance in NEI 07-13. In addition, proper identification and adequate description of these design features are required by 10 CFR 50.150. Therefore, NRC staff issued RAI 19-15 requesting the applicant to revise the DCD to indicate a 3-hour fire-rating for those 5psid doors identified under NEI 07-13. The staff and the applicant have had discussions dating back to January 27, 2010 in which the following DCD statement was discussed "…either a 5-psid door or two 3-hour rated fire doors…" The applicant should decide which option to be certified within the DCD. Proper identification and adequate description of these design features are required by 10 CFR 50.150. For example, Room 512 within the DCD shows access from corridor A via a vestibule. The applicant appeared to be replacing the non-rated fire door with a 3-hour fire-rated door without describing whether the appropriate vestibule walls also have 3-hour fire-rating. Therefore, the staff requested the applicant to revise the DCD by choosing one option for each situation. The applicant's response to RAI 19-15 dated June 21, 2010 (U7-C-STP-NRC-100152, ML101750070), states that Appendix 9A will be updated to include the 3-hour fire rating to all 5 psid doors required to be fire doors. In addition, the applicant will remove all "…either a 5-psid door or two 3-hour rated fire doors" statements within Appendix 9A. Each removed statement will be replaced with the actual design feature utilized within the applicant's assessment. In the response to RAI 19-15, the applicant referenced a proposed DCD markup in a letter dated June 17, 2010 (U7-C-STP-NRC-100139, ML101720306) that the staff found acceptable. The staff confirmed that the final DCD AIA Amendment provides the proposed DCD changes. Therefore, RAI 19-15 is resolved and closed.

Furthermore, the applicant included in the revised submittal dated May 12, 2010 (ML101340548) that the AFI [instrumentation] cabling itself will be 3-hour fire rated as it traverses through the reactor building and out to the AFI pump house. The staff accepts fire rated cabling, as opposed to cable wrapping, as this will ensure any AFI [instrumentation] cabling exposed to a fire-induced overpressure situation will still maintain the integrity of the fire protection rating. In the applicant's submittal of the amended ABWR DCD Revision 3, dated September 23, 2010 (ML102870017), the applicant added that the AFI instrumentation lines will be routed through rooms protected from fire exposure with the exception of Room 230. In this room, STPNOC states the AFI instrumentation line will be protected from fire utilizing a 3-hour fire rated wrap. The staff finds the routing of these instrumentation lines and limited use of fire rated wrap acceptable as the routing will ensure lines are not exposed to fire and the wrap is acceptable. This use of fire rated cable wrap is acceptable in this case because this room is not exposed to any overpressure but rather fire at ambient pressure only. The staff accepts that no other fire protection is required for the AFI system since no other portions of the AFI system which are vulnerable to fire reside within the reactor or control buildings.

These fire protection features protect the credited core cooling equipment described in 19S.4.2 above and protect the AFI instrumentation lines and cabling located in the reactor building. Based on the information in the application and the responses to the above RAIs, the staff found that the applicant has adequately described the fire protection key design features for maintaining core cooling in Sections 9.5.1, 9A and 19S.

*Primary Containment Structure*

In DCD Tier 2, Revision 3, Appendix 19S, the applicant states that the primary containment is a key design feature that will provide physical protection to the safety systems located inside the primary containment. NRC staff reviewed the information in DCD Tier 2, Revision 3 general arrangement drawings (Figure 1.2-1); containment and reactor building drawings (Figures 1.2-2, 1.2-2a, 1.2-8 through 1.2-12, 2.15.10j, and 2.15.10k); DCD Tier 2; and Sections 3.8 and 3H.1. The applicant states that the primary containment is entirely surrounded by the reactor building structure and therefore, a direct impact on the primary containment from a large commercial aircraft is not possible. The staff verified the applicant's statement by reviewing the cited drawings. The applicant determined by analysis that a strike on the primary containment (1) would not result in the perforation of the primary containment, and (2) would not cause direct damage to the systems within the containment or expose them to jet fuel.

Based on this review, the staff concluded that the applicant has adequately described the primary containment as a key design feature for protecting safety systems inside the primary containment to maintain core cooling.

*Reactor Building Structure*

In DCD Tier 2, Revision 3, Appendix 19S, the applicant states that the location and design of the reactor building structure are key design features that protect portions of the primary containment and the south wall of the control building from the impact of a large, commercial aircraft. This includes the protection provided by exterior walls, interior walls, intervening structures and barriers on the large openings in the reactor building exterior walls. The applicant states that ABWR containment is integrated with, and fully contained within, the reactor building. The containment and the reactor building are supported by a 5.5m thick common foundation mat. The bottom of the foundation mat is embedded in the ground approximately 26m below grade. Figure 1.2-1 shows the location of the reactor building in relation to other plant structures. The staff reviewed the information in DCD Tier 2, Revision 3 general arrangement drawings (Figures 1.2-1 1.2-2, 1.2-2a, 1.2-8 through 1.2-12 2.15.10j and 2.15.10k ), Section 3.8, and 3H.1, and verified the accuracy of the above description of the reactor building.

Based on the above review, the staff found adequate (in level of detail and scope) and acceptable the applicant's description of the reactor building and its design, as a key design feature for protecting portions of the primary containment and the south wall of the control building from the impact of a large, commercial aircraft.

*Turbine Building*

In DCD Tier 2, Revision 3, Appendix 19S, the applicant states that the location and design of the turbine building structure, as described in DCD Tier 1, Section 2.15.11 and shown in DCD Tier 2, Revision 3, general arrangement drawings Figure 1.2-1, and Figures 1.2-24 through 1.2-31 are key design features that protect portions of the north wall of the control and reactor building from the impact of a large, commercial aircraft. The turbine building is a Non-seismic Category I structure located to the north of the control building. The staff reviewed DCD Tier 2, Revision 3, general arrangement drawings, the above noted figures, Section 3.8.4 and Section 3H.2. The review led the staff to conclude that the above referenced drawings and documents contain acceptable level of detail and information to adequately describe the turbine building structure. Based on its review the staff found adequate and acceptable the applicant's description of the turbine building as a key design feature for protecting portions of the north wall of the control and reactor building from the impact of a large, commercial aircraft.

*Control Building Structures*

In DCD Tier 2, Revision 3, Appendix 19S, the applicant states that the location and design of the control building structure as shown in DCD Tier 2, Revision 3, (general arrangement drawings, Figure 1.2-1); Section 3.8.4; and Section 3H.2 are key design features that protect the reactor building from the impact of a large commercial aircraft. NRC staff reviewed the above information and control building-related figures to confirm the applicant's statement. Based on this review, the staff found adequate and acceptable the applicant's description of the control building as a key design feature for protecting portions of the north wall of the reactor building from the impact of a large commercial aircraft.

The staff found that the applicant has adequately described the key design features for providing physical protection to the reactor building by maintaining core cooling.

### 19S.4.4   Key Design Features that Maintain Integrity of the Spent Fuel Pool

The key design features credited to maintain the integrity of the spent fuel pool (SFP) are the location and design of the SFP and its support structure, as described in DCD Tier 2, Revision 3, Sections 1.2 and 9.1 and Figure 1.2-12. The applicant indicates that the location and design of the SFP and its support structure ensure that the SFP can withstand the effects of an impact of a large commercial aircraft. In the initial review of the DCD amendment (Tier 2, Appendix 19S), NRC staff noted that the applicant provided no statement supporting a conclusion that there would be no leakage from the SFP liner that would allow the SFP to drain below the required minimum water level, in accordance with the sufficiency criteria in NEI 07-13. The staff issued RAI 19-9 requesting the applicant to identify any scenarios resulting in SFP liner leakage and drainage below the required minimum water level. The applicant's response to RAI 19-9 in a letter dated February 18, 2010 (U7-C-STP-NRC-100040, ML100550027), clarifies that there are no impact scenarios that will result in leakage from the SFP liner below the required minimum water level. The staff issued RAI 19-10 requesting the applicant to describe, in the DCD, how the SFP location and its design protect the integrity of the SFP and prevent perforation below the required minimum water level. The applicant's response to

RAI 19-10 in a letter dated February 18, 2010 (ML100550027), clarifies that the SFP is located entirely within the reactor building and a detailed analysis of an aircraft impact showed that the structural design—which includes the SFP walls, liner, and support structures—is adequate to prevent a rupture of the liner. In addition, the applicant states that all pipes are configured so that they will not allow drainage below the minimum water level. The staff found that this clarification will allow the integrity of the SFP to be maintained. In the responses to both RAI 19-9 and RAI 19-10, the applicant provides a proposed DCD markup that the staff found acceptable. The staff confirmed that the revised DCD AIA Amendment (ML101340548) provides the proposed DCD changes; therefore, RAI 19-9 and RAI 19-10 are resolved. The staff found that the applicant has adequately described the key design features for ensuring the integrity of the SFP. The staff determined that the description of the key design features is in compliance with 10 CFR 50.150 and is therefore adequate and acceptable. Therefore, RAI 19-9 and RAI 19-10 are closed.

### 19S.4.5  Inspections, Tests, Analyses, and Acceptance Criteria (ITAAC)

During the initial review of DCD amendment submittal dated June 30, 2009 (ML092040048), NRC staff discovered the applicant failed to provide sufficient ITAAC to verify that the AFI system's key design characteristics and performance requirements are verified to ensure that the AFI system and its supporting systems will be available when required following an aircraft impact event.

The staff issued RAI 14.02-1 requesting the applicant to propose ITAAC for the AFI system that are necessary and sufficient to provide reasonable assurance that, if the inspections, tests, and analyses are performed and the acceptance criteria are met, a facility referencing the amended and certified ABWR design is constructed and will be operated in conformity with the design certification and NRC regulations. The applicant's response to RAI 14.02-1 dated April 8, 2010 (U7-C-STP-NRC-100049, ML101040254) provides several ITAAC for the AFI system in DCD Tier 1, Section 2.11.24 and Table 2.11.24:

**ITAAC# 2.11.24-1** states that the basic configuration of the AFI system is as described in Section 2.11.24. The staff found this response acceptable because Section 2.11.24 provides the high-level objective and design requirements of this system that the COL applicant must adhere to.

**ITAAC# 2.11.24-2** states that the AFI pump is capable of injecting ≥800 gpm into the RPV at the lowest SRV safety lift pressure. The staff found a minimum flow rate of 800 gpm to be an acceptable flow rate in order to maintain core cooling. The staff found the injection pressure of the lowest SRV safety lift pressure to be acceptable because that is the maximum pressure the AFI system is reasonably anticipated to be used for.

**ITAAC# 2.11.24-3** states that the AFI system water supply has a minimum capacity of 300,000 gallons and is refillable. This amount of water (300,000 gallons) is expected to provide 24 hours of core cooling capability, as discussed in Section 19.S.4.2 of DCD, which the staff found acceptable. A refillable water supply does not hinder but rather enhances the safety capabilities of the AFI system and therefore is an acceptable acceptance criterion.

**ITAAC# 2.11.24-4 through 2.11.24-6** state that the AFI pump house, AFI water supply, and AFI power supply (respectively) are located a minimum of 300 feet from the nearest outside wall of each building: the reactor building, control building, and turbine building. The staff found this description acceptable because the AFI pump house, which houses most of the

AFI equipment, together with the water supply and power supply are required to survive certain impact scenarios that render the ECCS inoperable. This separation distance is necessary so that both the target area and AFI pump house are not damaged by a single post-impact explosion. At least one of these systems is required in order to maintain core cooling.

**ITAAC# 2.11.24-7 through 2.11.24-9** state that barriers defined in NEI 07-13, Revision 7 exist between each AFI pump house, AFI water supply, and AFI power supply (respectively) and between each building: the reactor building, control building, and turbine building. The staff found this description acceptable because the AFI pump house, AFI water supply, and AFI power supply, together, are required to survive certain impact scenarios where the reactor building, control building, and/or turbine building have not. These intervening barriers defined in NEI 07-13, Revision 7 ensure that the physical impact of an aircraft does not destroy both the AFI system and the ECCS. At least one of these systems is required to survive in order to maintain core cooling.

**ITAAC# 2.11.24-10** states that instrumentation exists to provide information to the operator in the AFI pump house for reactor vessel water level, reactor pressure, suppression pool water level, and wetwell pressure. The staff found this description acceptable because this is the minimum information needed by an operator injecting water into the core in order to maintain core cooling.

**ITAAC# 2.11.24-11** states that MOVs in the AFI system injection line operate as designed on a manual initiation signal. The staff found this description acceptable because the AFI system is a manually operated system and therefore these MOVs, which are normally closed, will need to be opened upon the AFI operator's request for the system to be successful. There are no automatic signals required because this system is used solely for the purpose of aircraft impact events which the NRC classifies as a beyond-design-basis event.

The staff reviewed the information in the revised DCD Tier 1, Section 2.11.24 and Table 2.11.24 in accordance with the guidance in SRP Section 14.3. The staff found that the applicant's response to RAI 14.02-1 and the proposed ITAAC, as shown in DCD Tier 1, Section 2.11.24 and Table 2.11.24 meet the ITAAC requirements (10 CFR 52.47(b)(1)), and provide a necessary and sufficient basis for the staff to conclude that there is reasonable assurance that, if the inspections, tests, and analyses are performed and the acceptance criteria are met, a facility referencing the amended and certified ABWR design is constructed and will be operated in conformity with the design certification and applicable regulations, including 10 CFR 50.150. The staff also found that the aggregate of ITAAC representing the AFI system are comprehensive and encompass the necessary functions to ensure the AFI system is capable of performing its intended function following an aircraft impact event. Furthermore, the staff found that the ITAAC descriptions are objective, verifiable, and consistent with the Tier 2 information. Based on this review and a review of the selection methodology and criteria for development of the Tier 1, the staff concluded that the top-level design features and performance characteristics of the AFI system are appropriately described in Tier 1, Section 2.11.24 and Table 2.11.24 and this information is acceptable, therefore, the applicant's response is acceptable. The staff verified that the revised DCD AIA Amendment (ML101340548) includes the proposed ITAACs, and RAI 14.02-1 is resolved and closed.

## 19S.5    Conclusion

The NRC staff's review confirmed that the applicant has performed an aircraft impact assessment that is reasonably formulated to identify design features and functional capabilities to show, with reduced use of operator action, that the acceptance criteria in 10 CFR 50.150(a)(1) are met.  The staff found that the applicant has adequately described the key design features credited to meet 10 CFR 50.150, including descriptions of how the key design features show that the acceptance criteria in 10 CFR 50.150(a)(1) are met.  Therefore, the staff found that the applicant meets the applicable requirements of 10 CFR 50.150(b).

# 20 GENERIC ISSUES

## 20.5.1.3 Identification of Potential Design Improvements

### 20.5.1.3.1 Introduction

This section of the SER documents the evaluation of potential design improvements consistent with information provided in the ABWR DCD FSER (NUREG–1503). Neither the certified ABWR DCD, nor the AIA amendment, contains Section 20.5.

On June 30, 2009, STPNOC submitted an application to amend the Design Certification Rule for the U.S. ABWR, (STPNOC Letter No. U7-C-STP-NRC-090070, ML 092040048) to address the requirements of 10 CFR 50.150, the Commission's new aircraft impact rule (10 CFR 50.150). As part of this application, in ABWR DCD Revision 3, STPNOC is proposing to add a new AFI system and pump house in addition to other equipment to meet the requirements of the new aircraft impact rule. On November 10, 2009, STPNOC submitted a supplemental environmental report, titled, "Applicant's Supplemental Environmental Report-Amendment to ABWR Standard Design Certification," to address the requirements of 10 CFR 51.55(b), (STPNOC Letter No. U7-C-STP-NRC-090180, ADAMS Accession No. ML093170454).

### 20.5.1.3.2 Summary of Application

As part of the aircraft impact amendment that proposes to add a new AFI system and pump house in addition to other equipment, STPNOC submitted a supplemental environmental report to address the requirements of 10 CFR 51.55(b). In this report, the applicant evaluated the impacts of the design changes on the assessment of Severe Accident Mitigation Design Alternatives (SAMDAs).

### 20.5.1.3.3 Regulatory Basis

The regulatory basis for reviewing this section is 10 CFR 50.34(f). The information on the identification of potential cost-beneficial SAMDAs that supports this review was provided in the environmental report pursuant to 10 CFR 51.55(b).

### 20.5.1.3.4 Technical Evaluation

NRC staff reviewed the applicant's supplemental environmental report and checked the referenced certified ABWR DCD, and NUREG–1503, Supplement 1.

The numerical values and discussion in the ABWR DCD FSER (NUREG-1503), Section 20.5.1.3 was based on the values in Section 19P of the Standard Safety Analysis Report (SSAR). As part of the design certification rulemaking, General Electric (GE) updated SSAR Section 19P, but did not include it in the DCD. Instead, GE relocated this discussion to GE's "Technical Support Document (TSD) for the ABWR," Revision 1, December 1994, which was contained in an attachment to a letter from GE to the NRC dated December 21, 1994, (ADAMS Accession No. ML100210563). NRC staff reviewed the updated information in the TSD, and in FSER Supplement 1 Section 20.5.1.3, indicated that the conclusions in the final environmental assessment issued with the design certification rule will remain unchanged.

The applicant's environmental report evaluated the impacts of the DCD Revision 3 design changes on the assessment of SAMDAs, which were evaluated in the TSD. The applicant's review concludes that the design changes do not result in identification of any new SAMDAs that could become cost beneficial.

NRC staff reviewed the applicant's evaluations of the design changes and confirmed that the new designs will not result in a change to the ABWR probabilistic risk assessment (PRA) or DCD Chapter 19. Therefore, the staff concluded that the new design would not impact the original PRA, which provides input to the SAMDA evaluation, and therefore, would not change the conclusions reached in NUREG–1503 Supplement 1 and in the final environmental assessment issued with the ABWR design certification rule.

### 20.5.1.3.5    Conclusion

Based on the above technical evaluations, NRC staff concluded that the applicant has adequately addressed 10 CFR 50.34(f). The staff determined the proposed design changes would not affect the conclusions reached in NUREG–1503 Supplement 1, Section 20.5.1.3. The staff determined that there would be no adverse impacts from complying with the requirements for consideration of aircraft impacts on conclusions reached by the NRC in its review of the original ABWR design certification (NUREG-1503).

# APPENDIX A

# REFERENCES

**American Standards of Mechanical Engineers (ASME):**

— — — — —, ASME NQA-1-1994, "Quality Assurance Requirements for Nuclear Facility Applications."

**American National Standards Institute/American Nuclear Society (ANSI/ANS)**

— — — — —, ANSI/ANS 3.2-1999, "Administrative Controls and Quality Assurance for the Operational Phase of Nuclear Power Plants."

**Institute of Electrical and Electronic Engineers (IEEE) Standards:**

— — — — —, IEEE Std 336-1985, "IEEE Standard Installation, Inspection, and Testing Requirements for Power, Instrumentation, and Control Equipment at Nuclear Facilities."

— — — — —, IEEE Std 498-1985, "IEEE Standard Requirements for the Calibration and Control of Measuring and Test Equipment Used in Nuclear Facilities."

— — — — —, IEEE Std 603-1991, "IEEE Standard Criteria for Safety Systems for Nuclear Power Generating Stations."

— — — — —, IEEE Std 603-1991, Clause 5.6.3, "Independence Between Safety Systems and Other Systems."

— — — — —, IEEE Std 603-1991, Clause 6.3, "Interaction Between the Sense and Command Features and Other Systems."

**Nuclear Energy Institute (NEI)**

— — — — —, NEI 06-14, Revision 8, "Quality Assurance Program Description," May 2010.

— — — — —, NEI 07-13, Revision 7, "Methodology for Performing Aircraft Impact Assessments for New Plant Designs," May 2009.

**U.S. Code of Federal Regulations (CFR):**

— — — — —, Title 10 CFR Part 21, "Reporting of Defects and Noncompliance."

— — — — —, Title 10 CFR Part 50, "Domestic Licensing of Production and Utilization Facilities."

— — — — —, Title 10 CFR Part 50, Appendix A, "General Design Criteria for Nuclear Power Plants."

— — — — —, Title 10 CFR Part 50, Appendix B, "Quality Assurance Criteria for Nuclear Power Plants and Fuel Reprocessing Plants."

— — — — —, Title 10 CFR 50.34, "Contents of Applications; Technical Information."

—————, Title 10 CFR 50.36, "Technical Specifications."

—————, Title 10 CFR 50.48, "Fire Protection."

—————, Title 10 CFR 50.55, "Protection and Safety Systems."

—————, Title 10 CFR 50.70, "Inspections."

—————, Title 10 CFR 50.71, "Maintenance of Records, Making of Reports."

—————, Title 10 CFR 50.120, "Training and Qualification of Nuclear Power Plant Personnel."

—————, Title 10 CFR 50.150, "Aircraft Impact Assessment."

—————, Title 10 CFR 52.47, "Contents of Applications; Technical Information."

—————, Title 10 CFR 52.63, "Finality of Standard Design Certifications."

—————, Title 10 CFR 50.73, "Relationship to Other Subparts."

—————, Title 10 CFR 52.79, "Contents of Applications; Technical Information in Final Safety Analysis Report."

—————, Title 10 CFR Part 54, "Requirements for Renewal of Operating Licenses for Nuclear Power Plants."

—————, Title 10 CFR Part 52, Appendix A, "Design Certification Rule for the U.S. Advanced Boiling-Water Reactor."


## U.S. Nuclear Regulatory Commission (NRC)

## Generic Letter (GL)

—————, GL 85-06, "Quality Assurance Guidance for ATWS Equipment That Is Not Safety-Related," April 16, 1985.

—————, GL 88-18, "Plant Record Storage on Optical Disks," October 20, 1988.

—————, GL 89-02, "Actions to Improve the Detection of Counterfeit and Fraudulently Marketed Products," March 21, 2009.

—————, GL 91-05, "Licensee Commercial-Grade Procurement and Dedication Programs," April 9, 1991.

**NUREG-Series Reports:**

— — — —, NUREG–0800 "Standard Review Plan for the Review of Safety Analysis Reports for Nuclear Power Plants," June 1987.

— — — —, NUREG–1503, "Final Safety Evaluation Report Related to the Certification of the Advanced Boiling Water Reactor Design," July 1994.

— — — —, NUREG–1503, Supplement 1, "Final Safety Evaluation Report Related to the Certification of the Advanced Boiling Water Reactor Design," July 1997.

**Regulatory Guides:**

— — — —, RG 1.8, Revision 3, "Qualification and Training of Personnel for Nuclear Power Plants," May 2000.

— — — —, RG 1.26, Revision 4, "Quality Group Classifications and Standards for Water-, Steam-, and Radioactive-Waste-Containing Components of Nuclear Power Plants," March 2007.

— — — —, RG 1.28, Revision 3, "Quality Assurance Program Requirements (Design and Construction)," August 1985.

— — — —, RG 1.29, Revision 4, "Seismic Design Classification," March 2007.

— — — —, RG 1.33, Revision 2, "Quality Assurance Program Requirements," February 1978.

— — — —, RG 1.37, Revision 1, "Quality Assurance Requirements for Cleaning of Fluid Systems and Associated Components of Water-Cooled Nuclear Power Plants," March 2007.

— — — —, RG 1.68, Revision 3, "Initial Test Programs for Water-Cooled Nuclear Power Plants," March 2007.

— — — —, RG 1.70, Revision 3, "Standard Format and Content of Safety Analysis Reports for Nuclear Power Plants (LWR Edition)," November, 1978.

— — — —, RG 1.155, "Station Blackout," August 1988.

— — — —, RG 1.189, Revision 1, "Fire Protection for Nuclear Power Plants," April 2001.

— — — —, RG 1.206, "Combined License Applications for Nuclear Power Plants," July 2007.

**Draft Regulatory Guides:**

— — — — —, DG-1176 "Guidance for the Assessment of Beyond-Design-Basis Aircraft Impacts," July 2009.

**Regulatory Issue Summaries:**

— — — — —, RIS 2000-018, "Guidance on Managing Quality Assurance Records in Electronic Media," October 23, 2000.

**Atomic Energy Act of 1954**, as Amended, Public Law 102-486 (106 Stat 2943), October 24, 1992, Available at http://www.nrc.gov/reading-rm/doc-collections/nuregs/staff/sr0980/

**ADAMS Accession Numbers:**

ML100210563—Letter dated December 21, 1994, to Document Control Desk, Subject: NEPA/SAMDA Submittal for ABWR, with attachment titled, "Technical Support Document for the ABWR."

ML052360625—Memorandum dated August 26, 2005, from Dale, F. Thatcher, to L. Raghavan, Subject: Approval of Nuclear Management Company Change to Quality Assurance Report, NMC-1 (MC7585, MC7587, MC7588, MC7589, MC7590, MC7591, MC7592).

ML003693241—Letter dated March 20, 2000, to Mr. H. L. Sumner, Jr., Vice President—Nuclear Hatch Project, Southern Nuclear Operating Company, Inc. Subject: Edwin I. Hatch Nuclear Power Station, Units 1 And 2, Re: Approval of Relief Request RR-27, Third-Year Interval Inservice Inspection Program (TAC Nos. MA6163 and MA6164).

ML052710224—Letter dated September 28, 2005, to Mr. Gregg R. Overbeck, Senior Vice President, Nuclear Arizona Public Service Company. Subject: Palo Verde Nuclear Generating Station, Units 1, 2, and 3 - Approval of Change to Quality Assurance Program (Commercial-Grade Calibration Services) (TAC Nos. MC4402, MC4403, and MC4404).

ML070510300—Letter dated April 25, 2007, to Adrian P. Heymer, Senior Director, New Plant Deployment, Nuclear Generation Division, Nuclear Energy Institute. Subject: Final Safety Evaluation for Technical Report NEI 06-14, "Quality Assurance Program Description" (Project No. 689; TAC No. Md3406).

ML073440472—Letter dated December 19, 2007, to Mr. Keith Greenaway, President/CEO, ACLASS Accreditation Services. Subject: Reply to Your Letter Dated September 26, 2007, Seeking Agency Assistance in Accepting ACLASS Accreditation Services.

ML081140564—Letter dated April 22, 2008, to Mr. R. Douglas Leonard, Jr. Managing Director, Laboratory Accreditation Bureau. Subject: Reply to Your Letter Dated February 29, 2008, Seeking Assistance in Accepting Laboratory Accreditation Bureau.

ML081330253—Letter dated May 14, 2008, to Mr. Patrick V. McCullen, Vice President, International Accreditation Service, Inc. Subject: Reply to Your Letter Dated March 3, 2008, Seeking Assistance in Accepting International Accreditation Service, Inc.

ML092040048—Letter dated June 30, 2009, to Document Control Desk at NRC. Subject: South Texas Project Application to Amend the Design Certificate Rule for the U.S. Advanced Boiling-Water Reactor (ABWR).

ML092650695—Letter dated November 3, 2009, to Mr. Russell J. Bell, Director New Plant Licensing, Nuclear Generation Division, Nuclear Energy Institute. Subject: Final Safety Evaluation for Technical Report NEI 06-14, "Quality Assurance Program Description," Revision 7.

ML093170454—Letter dated November 10, 2009, to Document Control Desk at NRC, Subject: South Texas Project, Units 3 & 4, Submittal of "Applicant's Supplemental Environmental Report—Amendment to ABWR Standard Design Certification."

ML100190088—Letter dated January 13, 2010, to Document Control Desk at NRC. Subject: South Texas Project, Units 3 and 4 - Response to Request for Additional Information Question 06.02.04-1 Related to Application to Amend the ABWR DCD Amendment Part 2, Tier 2, Section 6.7 "High Pressure Nitrogen Gas Supply System" Provided in Attachment.

ML100210563—Letter dated December 21, 1994, to Document Control Desk at NRC. Subject: "NEPA/SAMDA Submittal for the ABWR." Letter from GE Nuclear Submitting Technical Support Document (TSD) for the ABWR.

ML100250139—Letter dated January 20, 2010, to Document Control Desk at NRC. Subject: South Texas Project, Units 3 and 4, Response to Request for Additional Information Letter Nos. 3 and 4 Related to the Application to Amend the ABWR DCD Part 2, Tier 2, Sections 5.2 and 7.7.

ML100470589—Letter dated February 8, 2010, to the Document Control Desk at the NRC. Subject: South Texas Project, Unit 3 and 4, Response to Request for Additional Information (RAI) Letter Number 10 Related to the Application to Amend the ABWR DCD Part 2, Tier 2, Section 1.0 Provided in Attachment 1 to the Referenced Letter.

ML100550025—Letter dated February 18, 2010, to Document Control Desk at NRC. Subject: South Texas Project, Unit 3 and 4, Transmittal of Response to Request for Additional Information (RAI) Letter Numbers 2 and 5 Related to the Application to Amend the ABWR DCD Part 2, Tier 2, Section 19, Provided in Attachment 1 to the Referenced Letter.

ML100550027—Letter dated February 18, 2010, to Document Control Desk at NRC. Subject: South Texas Project, Unit 3 and 4, Transmittal of Response to Request for Additional Information (RAI) Letter Numbers 5, 6, 7, and 8, Related to the Application to Amend the ABWR DCD Part 2, Tier 2, Sections 3.8, 5.2, 15.8, and 19, Provided in Attachment 1 to the Referenced Letter.

ML100600410—Letter dated February 25, 2010, to Document Control Desk at NRC. Subject: South Texas Project, Units 3 and 4, Response to Request for Additional Information RAI Letter Numbers 5, 8, and 9 Related to the Application to Amend the ABWR DCD Part 2, Tier 2, Sections 3.8, 9.2, and 19.

ML100640162—Letter dated March 3, 2010, to Document Control Desk at NRC. Subject: South Texas Project, Units 3 and 4 - Response to Request for Additional Information.

ML100770388—Letter dated March 17, 2010, to Document Control Desk at NRC. Subject: South Texas Project, Units 3 and 4 - Responses to Request for Additional Information.

ML101040254—Letter dated April 8, 2010, to Document Control Desk at NRC. Subject: South Texas Project, Units 3 and 4—Responses to Request for Additional Information, Letter Number 11 Related to the Application to Amend the ABWR DCD Part 2, Tier 2, Section 14.2.

ML101040345—Letter dated April 8, 2010, to Document Control Desk at NRC. Subject: South Texas Project, Units 3 and 4—Responses to Request for Additional Information, Related to the Application to Amend the ABWR DCD Part 2, Tier 2, Section 1.0

ML101120085—Letter dated April 19, 2010, to Document Control Desk at NRC. Subject: South Texas Project, Units 3 and 4—Responses to Request for Additional Information Question 09.02.04-1.

ML101190120—Letter dated April 26, 2010, to Document Control Desk at NRC. Subject: South Texas Project, Units 3 and 4 - Response to Request for Additional Information Question 03.02.02-1 Related to the Application to Amend the ABWR DCD Part 2, Tier 2, Section 3.2.

ML101340548—Letter dated May 12, 2010, to Document Control Desk at NRC. Subject: South Texas Project Proposed ABWR DCD AIA Amendment Revision 1.

ML101470298—Draft Letter from NRC to Mr. Scott M Head, South Texas Project Nuclear Operating Company, Subject: NRC Inspection Report Nos. 05200012/2010201 and 5200013/2010201, dated July 21, 2010.

ML101530610—Letter dated May 27, 2010, to Document Control Desk at NRC. Subject: South Texas Project Units 3 and 4—Responses to Request for Additional Information Questions 03.08.04-2 and 03.08.04-4 Related to the Application to Amend the ABWR DCD Part 2, Tier 2, Section 3.2.

ML101720306—Letter dated June 17, 2010, to Document Control Desk at NRC. Subject: STP ABWR Aircraft Impact Assessment (AIA) Revision.

ML101470298—Letter dated June 21, 2010, to Document Control Desk at NRC. Subject: South Texas Project Units 3 and 4—Responses to Request for Additional Information Letter Number 13 Related to the Application to Amend the ABWR DCD Part 2, Tier 2, Section 19.

ML102000496—Letter dated July 12, 2010, to Document Control Desk at NRC. Subject: ABWR STP Aircraft Impact Assessment (AIA) Amendment Revision-0.

ML102240435—Letter dated August 4, 2010, to Document Control Desk at NRC. Subject: ABWR STP Aircraft Impact Assessment (AIA) Amendment Revision-1.

ML103190120—Letter dated September 2, 2010, to Document Control Desk at NRC. Subject: ABWR STP Aircraft Impact Assessment (AIA) Amendment Revision-2.

ML102870017—Letter dated September 23, 2010, to Document Control Desk at NRC. Subject: ABWR STP Aircraft Impact Assessment (AIA) Amendment Revision-3.

# APPENDIX B

# CHRONOLOGY OF CORRESPONDENCE

| Document Date | ADAMS Accession Number | Subject | Correspondence | From | To | Docket Number |
|---|---|---|---|---|---|---|
| 6/30/2009 | ML092040048 | South Texas Project Application to Amend the Design Certificate Rule for the US Advanced Boiling Water Reactor (ABWR). | Letter | South Texas Project Nuclear Operating Company | NRC/Document Control Desk NRC/NRO | 05200001 |
| 11/10/2009 | ML093170454 | South Texas Project, Units 3 & 4, Submittal of "Applicant's Supplemental Environmental Report-Amendment to ABWR Standard Design Certification." | Letter | South Texas Project Nuclear Operating Company | NRC/Document Control Desk NRC/NRO | 05200001 |
| 11/10/2009 | ML093170511 | Transmittal Letter of 10/14/2009 Meeting Presentation Materials. | Letter Meeting Briefing Package/ Handouts Slides and Viewgraphs | South Texas Project Nuclear Operating Company | NRC/Document Control Desk NRC/NRO | 05200001 |
| 1/13/2010 | ML100190088 | South Texas Project, Units 3 & 4 - Response to Request for Additional Information Question 06.02.04-1 Related to Application to Amend the ABWR DCD Amendment Part 2, Tier 2, Section 6.7 "High Pressure Nitrogen Gas Supply System" Provided in Attachment. | Letter | South Texas Project Nuclear Operating Company | NRC/Document Control Desk NRC/NRO | 05200001 |
| 1/20/2010 | ML100250139 | South Texas Project, Units 3 and 4, Response to Request for Additional Information Letter Nos. 3 and 4 Related to the Application to Amend the ABWR DCD Part 2, Tier 2, Sections 5.2 and 7.7. | Letter | South Texas Project Nuclear Operating Company | NRC/Document Control Desk NRC/NRO | 05200001 |
| 2/8/2010 | ML100470589 | South Texas Project, Unit 3 & 4, Response to Request for Additional Information (RAI) Letter Number 10 Related to the Application to Amend the ABWR DCD Part 2, Tier 2, Section 1.0 Provided in Attachment 1 to the Referenced Letter. | Letter | South Texas Project Nuclear Operating Company | NRC/Document Control Desk NRC/NRO | 05200001 |
| 2/18/2010 | ML100550025 | South Texas Project, Unit 3 & 4, Transmittal of Response to Request for Additional Information (RAI) Letter Numbers 2 and 5 Related to the Application to Amend the ABWR DCD Part 2, Tier 2, Section 19 Provided in Attachment 1 to the Referenced Letter. | Letter | South Texas Project Nuclear Operating Company | NRC/Document Control Desk NRC/NRO | 05200001 |
| 2/18/2010 | ML100550027 | South Texas Project, Units 3 & 4, Response to Request for Additional Information Letter No. 5, 6, 7, & 8, Related to the Application to Amend the ABWR DCD Part 2, Tier 2, Sections 3.8, 5.2, 15.8 and 19. | Letter | South Texas Project Nuclear Operating Company | NRC/Document Control Desk NRC/NRO | 05200001 |

| Document Date | ADAMS Accession Number | Subject | Correspondence | From | To | Docket Number |
|---|---|---|---|---|---|---|
| 2/25/2010 | ML100600410 | South Texas Project, Units 3 & 4, Response to Request for Additional Information RAI Letter Numbers 5, 8, and 9 Related to the Application to Amend the ABWR DCD Part 2, Tier 2, Sections 3.8, 9.2, and 19. | Letter | South Texas Project Nuclear Operating Company | NRC/Document Control Desk NRC/NRO | 05200001 |
| 4/8/2010 | ML101040254 | South Texas Project, Units 3 & 4, Response to Request for Additional Information Letter Number 11 Related to the Application to Amend the ABWR DCD Part 2, Tier 2, Section 14.2. | Letter | South Texas Project Nuclear Operating Company | NRC/Document Control Desk NRC/NRO | 05200001 |
| 4/8/2010 | ML101040345 | South Texas Project, Units 3 & 4, Response to Request for Additional Information Related to Application to Amend ABWR DCD Part 2, Tier 2, Section 1.0 provided. | Letter | South Texas Project Nuclear Operating Company | NRC/Document Control Desk NRC/NRO | 05200001 |
| 4/19/2010 | ML101120085 | South Texas Project, Units 3 and 4, Response to Request for Additional Information Question 09.02.04-1. | Letter | South Texas Project Nuclear Operating Company | NRC/Document Control Desk NRC/NRO | 05200001 |
| 4/26/2010 | ML101190120 | South Texas Project, Units 3 and 4, Response to Request for Additional Information Question 03.02.02-1 Related to Application Amend ABWR DCD, Part 2, Tier 2, Section 3.2. | Letter | South Texas Project Nuclear Operating Company | NRC/Document Control Desk NRC/NRO | 05200001 |
| 5/12/2010 | ML101340548 | South Texas Project, Units 3 and 4, Proposed ABWR DCD AIA Ammendment Revision 1 | Letter | South Texas Project Nuclear Operating Company | NRC/Document Control Desk NRC/NRO | 05200001 |
| 7/21/2010 | ML101470298 | NRC Inspection Report Nos. 05200012/2010201 and 05200013/2010201 | Letter with Enclosures | NRC, Quality and Vendor Branch 2, Office of New Reactor | South Texas Project Nuclear Operating Company | 05200012 05200013 |
| 5/27/2010 | ML101530610 | South Texas Project, Units 3 and 4, Response to Request for Additional Information Question 03.08.04-2 and 03.08.04-4 Related to Application to Amend ABWR DCD, Part 2, Tier 2, Section 3.2. | Letter | South Texas Project Nuclear Operating Company | NRC/Document Control Desk NRC/NRO | 05200001 |
| 6/17/2010 | ML101720306 | South Texas Project, Units 3 and 4, STP ABWR DCD Aircraft Impact Assessment (AIA) Ammendment Revision | Letter | South Texas Project Nuclear Operating Company | NRC/Document Control Desk NRC/NRO | 05200001 |
| 6/21/2010 | ML101750070 | South Texas Project, Units 3 and 4, Response to Request for Additional Information Letter Number 13 Related to Application to Ammend ABWR DCD, Part 2, Tier 2, Section 19. | Letter | South Texas Project Nuclear Operating Company | NRC/Document Control Desk NRC/NRO | 05200001 |
| 7/12/2010 | ML102000496 | South Texas Project, Units 3 and 4, STP ABWR DCD Aircraft Impact Assessment (AIA) Ammendment Revision-0 | Letter | South Texas Project Nuclear Operating Company | NRC/Document Control Desk NRC/NRO | 05200001 |

| Document Date | ADAMS Accession Number | Subject | Correspondence | From | To | Docket Number |
|---|---|---|---|---|---|---|
| 8/4/2010 | ML102240435 | South Texas Project, Units 3 and 4, STP ABWR DCD Aircraft Impact Assessment (AIA) Ammendment Revision-1 | Letter | South Texas Project Nuclear Operating Company | NRC/Document Control Desk NRC/NRO | 05200001 |
| 9/2/2010 | ML103190120 | South Texas Project, Units 3 and 4, STP ABWR DCD Aircraft Impact Assessment (AIA) Ammendment Revision-2 | Letter | South Texas Project Nuclear Operating Company | NRC/Document Control Desk NRC/NRO | 05200001 |
| 9/23/2010 | ML102870017 | South Texas Project, Units 3 and 4, STP ABWR DCD Aircraft Impact Assessment (AIA) Ammendment Revision-3 | Letter | South Texas Project Nuclear Operating Company | NRC/Document Control Desk NRC/NRO | 05200001 |

# APPENDIX C

## PRINCIPAL CONTRIBUTORS

| | |
|---|---|
| Paul Kallan | Senior Project Manager |
| Nanette Gilles | Senior Policy Analyst (AIA Working Group) |
| Mark Caruso | Senior Reliability and Risk Engineer (AIA Working Group) |
| Bhagwat Jain | Senior Structural Engineer (AIA Working Group) |
| Dennis Andrukat | General Engineer (AIA Working Group) |
| David Jeng | Senior Structural Engineer (AIA Working Group) |
| Raj Goel | Reactor Systems Engineer |
| George Thomas | Senior Reactor Engineer |
| James Gilmer | Reactor Systems Engineer |
| Erick Martinez | General Engineer |
| Dinesh Taneja | Senior Electronics Engineer |
| Frank Talbot | Operations Engineer |
| Garrett Newman | Reactor Operations Engineer |
| Richard McIntyre | Senior Reactor Engineer |
| Angelo Stubbs | Senior Reactor Systems Engineer |
| Edward Fuller | Senior Reliability and Risk Engineer |
| Richard McNally | Mechanical Engineer |
| Mark Lintz | Operations Engineer |
| Peyton Doub | Environmental Scientist |
| Frank Akstulewicz | Management Supervision |
| Mark Tonacci | Management Supervision |
| Kimberly Hawkins | Management Supervision |
| Brent Clayton | Management Supervision |
| Lynn Mrowca | Management Supervision |
| John Segala | Management Supervision |
| John McKirgan | Management Supervision |
| Ian Jung | Management Supervision |
| Joseph Donoghue | Management Supervision |
| Richard Rasmussen | Management Supervision |
| Brian Thomas | Management Supervision |
| Jason Dreisbach | Management Supervision |
| Michael Junge | Management Supervision |
| Jennifer Dixon-Herrity | Management Supervision |
| Jerry Wilson | Peer Review |
| Tom Tai | Peer Review |
| Bernadette Abeywickrama | Licensing Assistant |
| Stacy Joseph | Project Manager |

Contractor

| | |
|---|---|
| Energy Research, Inc. | SER Support |

# APPENDIX D

# CHRONOLOGY OF REQUESTS FOR ADDITIONAL INFORMATION (RAI)

| Question Number: | SRP Section Title: | RAI Issued: | RAI Accession No: | RAI Response: | Response Accession No: |
|---|---|---|---|---|---|
| 01-2 | 01 - Introduction and Interfaces | 1/13/10 | ML100130817 | 2/8/10 | ML100470589 |
| 03.02.02-1 | 03.02.02 - System Quality Group Classification | 1/26/10 | ML100320305 | 3/3/10 | ML100640162 |
| 03.08.04-1 | 03.08.04 - Other Seismic Category I Structures | 1/4/10 | ML100050150 | 2/18/10 | ML100550027 |
| 03.08.04-2 | 03.08.04 - Other Seismic Category I Structures | 1/4/10 | ML100050150 | 5/27/10 | ML101530610 |
| 03.08.04-3 | 03.08.04 - Other Seismic Category I Structures | 1/4/10 | ML100050150 | 2/18/10 | ML100550027 |
| 03.08.04-4 | 03.08.04 - Other Seismic Category I Structures | 1/4/10 | ML100050150 | 5/27/10 | ML101530610 |
| 03.08.04-5 | 03.08.04 - Other Seismic Category I Structures | 1/4/10 | ML100050150 | 2/18/10 | ML100550027 |
| 05.02.02-1 | 05.02.02 - Overpressure Protection | 12/18/09 | ML093560920 | 1/20/10 | ML100250139 |
| 05.02.02-2 | 05.02.02 - Overpressure Protection | 12/18/09 | ML093560920 | 1/20/10 | ML100250139 |
| 05.02.02-3 | 05.02.02 - Overpressure Protection | 1/4/10 | ML100050050 | 2/18/10 | ML100550027 |
| 06.02.04-1 | 06.02.04 - Containment Isolation System | 12/16/09 | ML093500126 | 1/13/10 | ML100190088 |
| 07.07-1 | 07.07 - Control Systems | 12/18/09 | ML093560881 | 1/20/10 | ML100250139 |
| 07.07-2 | 07.07 - Control Systems | 12/18/09 | ML093560881 | 1/20/10 | ML100250139 |
| 07.07-3 | 07.07 - Control Systems | 12/18/09 | ML093560881 | 1/20/10 | ML100250139 |
| 09.02.04-1 | 09.02.04 - Potable and Sanitary Water Systems | 1/13/10 | ML100140773 | 4/26/10 | ML101120085 |
| 09.02.04-2 | 09.02.04 - Potable and Sanitary Water Systems | 1/13/10 | ML100140773 | 2/25/10 | ML100600410 |
| 09.02.04-3 | 09.02.04 - Potable and Sanitary Water Systems | 1/13/10 | ML100140773 | 2/25/10 | ML100600410 |
| 09.02.04-4 | 09.02.04 - Potable and Sanitary Water Systems | 1/13/10 | ML100140773 | 2/25/10 | ML100600410 |
| 14.02-1 | 14.02 - Initial Plant Test Program - Design Certification and New License Applicants | 1/14/10 | ML100190138 | 8/4/10 | ML101040254 |
| 15.08-1 | 15.08 - Anticipated Transients Without Scram | 12/30/09 | ML093641061 | 2/18/10 | ML100550027 |
| 19-1 | 19 - Probabilistic Risk Assessment and Severe Accident Evaluation | 12/18/09 | ML093520830 | 2/18/10 | ML100550025 |
| 19-2 | 19 - Probabilistic Risk Assessment and Severe Accident Evaluation | 12/18/09 | ML093520830 | 2/18/10 | ML100550025 |
| 19-3 | 19 - Probabilistic Risk Assessment and Severe Accident Evaluation | 12/18/09 | ML093520830 | 2/18/10 | ML100550025 |
| 19-4 | 19 - Probabilistic Risk Assessment and Severe Accident Evaluation | 12/18/09 | ML093520830 | 2/18/10 | ML100550025 |

| Question Number: | SRP Section Title: | RAI Issued: | RAI Accession No: | RAI Response: | Response Accession No: |
|---|---|---|---|---|---|
| 19-5 | 19 - Probabilistic Risk Assessment and Severe Accident Evaluation | 12/18/09 | ML093520830 | 2/18/10 | ML100550025 |
| 19-6 | 19 - Probabilistic Risk Assessment and Severe Accident Evaluation | 12/18/09 | ML093520830 | 2/18/10 | ML100550025 |
| 19-7 | 19 - Probabilistic Risk Assessment and Severe Accident Evaluation | 12/18/09 | ML093520830 | 2/18/10 | ML100550025 |
| 19-8 | 19 - Probabilistic Risk Assessment and Severe Accident Evaluation | 12/22/09 | ML093641050 | 2/18/10 | ML100550027 |
| 19-9 | 19 - Probabilistic Risk Assessment and Severe Accident Evaluation | 12/22/09 | ML093641050 | 2/18/10 | ML100550027 |
| 19-10 | 19 - Probabilistic Risk Assessment and Severe Accident Evaluation | 12/22/09 | ML093641050 | 2/18/10 | ML100550027 |
| 19-11 | 19 - Probabilistic Risk Assessment and Severe Accident Evaluation | 12/22/09 | ML093641050 | 2/25/10 | ML100600410 |
| 19-12 | 19 - Probabilistic Risk Assessment and Severe Accident Evaluation | 12/22/09 | ML093641050 | 2/18/10 | ML100550027 |
| 19-13 | 19 - Probabilistic Risk Assessment and Severe Accident Evaluation | 12/22/09 | ML093641050 | 2/18/10 | ML100550025 |
| 19-14 | 19 - Probabilistic Risk Assessment and Severe Accident Evaluation | 5/20/10 | ML101410197 | 6/21/10 | ML101750070 |
| 19-15 | 19 - Probabilistic Risk Assessment and Severe Accident Evaluation | 5/20/10 | ML101410197 | 6/21/10 | ML101750070 |

# APPENDIX E

# REPORT BY ADVISORY COMMITTEE ON REACTOR SAFEGUARDS

**UNITED STATES
NUCLEAR REGULATORY COMMISSION
ADVISORY COMMITTEE ON REACTOR SAFEGUARDS
WASHINGTON, DC 20555 - 0001**

September 20, 2010

The Honorable Gregory B. Jaczko
Chairman
U.S. Nuclear Regulatory Commission
Washington, DC 20555-0001

SUBJECT: REPORT ON THE SAFETY ASPECTS OF THE SOUTH TEXAS PROJECT
NUCLEAR OPERATING COMPANY APPLICATION TO AMEND THE
CERTIFIED U.S. ABWR DESIGN TO INCORPORATE THE AIRCRAFT IMPACT
ASSESSMENT RULE

Dear Chairman Jaczko:

During the 575th meeting of the Advisory Committee on Reactor Safeguards, September 9-11, 2010, we reviewed the staff's Safety Evaluation Report (SER) related to the South Texas Project Nuclear Operating Company (STPNOC) application to amend the certified U.S. Advanced Boiling Water Reactor (ABWR) design. The purpose of the amendment is to address the requirements of the Aircraft Impact Assessment (AIA) Rule specified in 10 CFR 50.150. Our ABWR Subcommittee held a meeting on August 18, 2010, to review the application, the AIA performed by the applicant, and the staff's SER and AIA inspection report. During these meetings, we had the benefit of discussions with representatives of the NRC staff and STPNOC and their supporting contractors. We also had the benefit of the documents referenced. This report fulfills the requirement of 10 CFR 52.53 that the ACRS report on those portions of the application which concern safety.

## CONCLUSIONS AND RECOMMENDATIONS

1. The STPNOC application to amend the ABWR design certification rule and the staff's SER are acceptable subject to satisfactory closure of the issues identified in the Notice of Violation and Recommendation 2.

2. The staff should ensure that the applicant demonstrates and documents that the temperature within the fire-protected area where the Alternate Feedwater Injection (AFI) system instrument rack is to be located will not exceed the instruments' environmental qualification conditions.

3. The staff should ensure that the assumptions and initial conditions credited in the applicant's AIA are properly incorporated into the amended Design Control Document (DCD).

4. The staff should ensure that Combined License (COL) applicants referencing this amendment have an appropriate process to assure the reliability of the AFI system.

5. The staff should complete a lessons-learned review of this application to identify any deficiencies in the AIA Inspection Procedure (IP) 37804 and the AIA methodology prescribed in Nuclear Energy Institute (NEI) 07-13, Revision 7. Resolution of these deficiencies should be communicated to the industry and incorporated into the staff's future reviews.

## BACKGROUND

The ABWR design was certified by the NRC on May 12, 1997. On June 30, 2009, STPNOC submitted an application to amend the ABWR design certification rule to address the requirements of the AIA rule specified in 10 CFR 50.150. The application was revised on May 12, 2010, and June 17, 2010. After incorporating responses to the NRC requests for additional information, STPNOC submitted the final application on July 12, 2010, which is the basis for this review and the staff's SER. Subsequent to the issuance of the SER, STPNOC submitted other revisions of their application to address deficiencies identified in the staff's inspection report and Notice of Violation.

As required by 10 CFR 50.150, applicants for new nuclear power plants must perform an assessment of the effects of the impact of a large, commercial aircraft on the designed facility. Using realistic analyses, applicants must identify and incorporate into the design those design features and functional capabilities needed to show that, with reduced use of operator action, (1) the reactor core remains cooled or the containment remains intact and (2) spent fuel cooling or spent fuel pool integrity is maintained (referred to as the acceptance criteria).

Applicants are required to submit a description of the design features and functional capabilities identified as a result of the AIA and a description of how those features and capabilities show that the acceptance criteria are met. Since the impact of a large, commercial aircraft is a beyond-design-basis event, applicants may use non-safety-related features or capabilities to satisfy the requirements of 10 CFR 50.150.

The Statement of Considerations for 10 CFR 50.150 states that the COL applicant is not required to submit the AIA to the NRC, but the assessment will be subject to inspection by the NRC, and must, therefore, be maintained by the applicant. On May 17-21, 2010, the staff conducted an inspection of the AIA at the applicant's supporting contractor facility. The inspection was performed in accordance with IP 37804, "Aircraft Impact Assessment," dated April 27, 2010. The AIA was made available to us by the applicant for review prior to our ABWR Subcommittee meeting of August 18, 2010.

If the proposed amendment is approved, applicants for a COL that reference the ABWR standard design can address the requirements of 10 CFR 50.150 by referencing the amended ABWR standard design.

## DISCUSSION

The certified ABWR design in Appendix A of 10 CFR Part 52 does not address the AIA rule as it predates the rule. The AIA performed by the applicant uses, without exceptions, the industry guidance in NEI 07-13, Revision 7, endorsed in Draft Regulatory Guide DG-1176. The results of the AIA show that the modified ABWR design described in the STPNOC application meets the acceptance criteria of the AIA rule by maintaining core cooling and integrity of the spent fuel pool.

The key design features identified by STPNOC to satisfy the requirements of 10 CFR 50.150 include: the primary containment structure; arrangement and design of the control, turbine, and reactor building structures; design and location of the spent fuel pool; physical separation of Class 1E emergency diesel generators; pressure-rated and non-pressure rated fire barriers; physical separation and design of the emergency core cooling systems; design of the containment overpressure protection system; and the AFI system.

Guided by the results of the AIA, the changes made to the certified ABWR design include: the addition of the AFI system and associated auxiliaries; addition of barriers at six large openings in the reactor building; addition or upgrades to 16 fire/watertight doors in the reactor building; definition of the structural characteristics of exterior and four interior walls on two floors in the reactor building; replacement of two gratings with concrete hatches; and enhancement of all three-hour fire barriers, including piping penetration seals, to a rating of 5 psid. The staff's review concluded that these changes will have no adverse impacts on conclusions reached in their review of the original ABWR design certification. We agree.

The AFI system provides an alternate means of water injection in the reactor pressure vessel (RPV). The AFI system has a dedicated high pressure pump sized to remove decay heat based on scram from full power. The AFI pump is installed in and can be started from a pump room located outside the strike-affected zone. The water source for the AFI system is also located outside the strike-affected zone. The water supply capacity for the AFI system is sufficient to remove decay heat by evaporation for nearly 24 hours. Piping from the AFI system is routed underground and is protected from damage by the aircraft impact. Water is injected into the feedwater lines and the tie-in points are in the steam tunnel. A reliable AC power supply, not affected by the aircraft impact, is to be provided for the pump. Additionally, a dedicated DC power supply is provided in the AFI pump room.

Since existing instrumentation may fail due to fires in the reactor building and control building, a shock-mounted AFI instrument rack has been added in a fire-protected area of the reactor building. A minimal set of instrumentation with a read-out in the AFI pump room is provided to allow the AFI system operator to monitor the RPV water level, RPV pressure, containment pressure, and suppression pool level. The instrument lines from the primary containment penetrations to the instrument rack are protected from plane wreckage and other debris by barriers. Power to the rack is provided from the AFI pump room so it is also protected from an aircraft impact.

[

] A limited set of operator actions is required to start the AFI system and control the RPV water level from the AFI pump room.

The analyses presented by the applicant demonstrate that for all of the assumed aircraft strike locations, core cooling will be maintained. A detailed structural analysis shows that even when limited credit is taken for the reactor building walls, the primary containment walls will not be perforated. [

] For the other strike scenarios, the AFI system is required to provide adequate core cooling. Analyses show that by placing the AFI system in service within 30 minutes of event initiation, the RPV water level can be maintained well above the top of active fuel. Since the core does not uncover, no fuel damage is expected. Removal of decay heat is accomplished by discharging steam through the RPV safety relief valves (SRVs) to the suppression pool. As the suppression pool gradually heats up, the primary containment pressure rises reaching the Containment Overpressure Protection System (COPS) setpoint (90 psig) within 20 hours of event initiation. The COPS automatically relieves the containment overpressure by venting the wetwell. The COPS relief capacity is well in excess of the core boil-off rate when the COPS setpoint is reached.

The spent fuel pool is located entirely within the reactor building. All pipes are configured to prevent drainage below the minimum water level. Structural analyses performed by the applicant show that for all of the assumed strike scenarios, without accounting for the reactor building exterior walls, the spent fuel pool liner will not be perforated, and the integrity of the spent fuel pool and its support structures will be maintained.

Based on the analyses performed by the applicant, we conclude that the enhanced ABWR design described in the STPNOC application will meet the acceptance criteria specified in 10 CFR 50.150 by maintaining core cooling and spent fuel pool integrity. However, several assumptions were made in the analyses because of insufficient design detail. For example, the original ABWR DCD does not specify routing of electrical cables. However, the AIA assumes that cable routing maintains divisional separation thereby limiting the damage resulting from different strike scenarios. It is therefore necessary to ensure that the assumptions and initial conditions credited in the applicant's AIA are properly incorporated into the amended DCD.

The AFI instrument rack is located in a fire-protected area of the reactor building. Guided by the results of the AIA, fire doors have been upgraded to prevent the spread of fire in that area. However, the AIA does not include an analysis to estimate the maximum temperature within that area, if it were to be surrounded by fire. It is necessary that such an analysis be performed and documented in the AIA to demonstrate that the conditions in which the AFI instruments may operate following an aircraft impact are consistent with their environmental qualification conditions.

The impact of a large, commercial aircraft is a beyond-design-basis event so that both safety and non-safety related structures, systems and components may be credited in the realistic assessment to address the AIA rule. The AFI system specified in the amendment is a non-safety system; all tie-ins to the AFI system are also non-safety related. There are no technical specifications associated with the AFI system. Considering the importance of the AFI system
[                                                                                          ]
the system must be reliable. The staff should ensure that COL applicants referencing this amendment have an appropriate process to assure the reliability of the AFI system.

The staff performed a thorough inspection of the applicant's AIA. The same headquarters personnel involved in the review of the application were also involved in the inspection. Maintaining the same personnel with the high level skill sets in the review of the application and the inspection significantly enhances the quality of the review process. We commend the staff for this action.

The staff's inspection report noted that NEI 07-13, Revision 7 is silent on specific characteristics of the fire barrier configuration such as the separation distance and pressure dissipation. Additionally, NEI 07-13, Revision 7 discusses the effect of an impact on the polar crane but not the gantry crane. These were among the deficiencies identified in the Notice of Violation. The staff's inspection report also noted that NEI 07-13, Revision 7 was not intended to be an all inclusive instruction and the fact that it is silent in some areas cannot be considered justification for excluding those areas from the scope of the AIA. Therefore, it is important for the staff to identify any deficiencies in IP 37804 and the AIA methodology prescribed in NEI 07-13, Revision 7. The staff has initiated a lessons-learned review of this application to identify such deficiencies. This review should be completed in a timely fashion so that the resolution of such deficiencies can be communicated to the industry and incorporated into the staff's future reviews.

The STPNOC application to amend the ABWR design certification rule and the staff's SER are acceptable subject to satisfactory closure of issues related to the Notice of Violation and our Recommendation 2.

Sincerely,

/RA/

Said Abdel-Khalik
Chairman

References:

1. Memorandum to Edwin M. Hackett, Advanced Final Safety Evaluation Report Regarding the Advanced Boiling Water Reactor Design Certification Review, 07/19/2010 (ML102000599)

2. Letter to Scott M. Head, South Texas Project Nuclear Operating Company Aircraft Impact Assessment Inspection, NRC Inspection Report No. 05200001/2010-202 and Notice of Violation, 08/13/2010 (ML102100218)

3. Federal Register Notice (74 FR 28112), "Aircraft Impact Assessment Rule Consideration of Aircraft Impacts for New Nuclear Power Reactors," 06/12/2009

4. Draft Regulatory Guide DG 1176, "Guidance for the Assessment of Beyond-Design-Basis Aircraft Impacts," 07/31/2009 (ML073170252)

5. NEI 07-13, Revision 7, "Methodology for Performing Aircraft Impact Assessments for New Plant Designs," 05/31/2009 (ML093570239)

6. Inspection procedure IP 37804, "Aircraft Impact Assessment," 04/27/2010 (ML100680701)

7. Letter to Chairman Jaczko, Draft Final Regulatory Guide 1.217, "Guidance for the Assessment of Beyond-Design-Basis Aircraft Impacts," 02/18/2010 (ML100470861)

Revision 7. The staff has initiated a lessons-learned review of this application to identify such deficiencies. This review should be completed in a timely fashion so that the resolution of such deficiencies can be communicated to the industry and incorporated into the staff's future reviews.

The STPNOC application to amend the ABWR design certification rule and the staff's SER are acceptable subject to satisfactory closure of issues related to the Notice of Violation and our Recommendation 2.

Sincerely,
/RA/
Said Abdel-Khalik
Chairman

References:

1. Memorandum to Edwin M. Hackett, Advanced Final Safety Evaluation Report Regarding the Advanced Boiling Water Reactor Design Certification Review, 07/19/2010 (ML102000599)

2. Letter to Scott M. Head, South Texas Project Nuclear Operating Company Aircraft Impact Assessment Inspection, NRC Inspection Report No. 05200001/2010-202 and Notice of Violation, 08/13/2010 (ML102100218)

3. Federal Register Notice (74 FR 28112), "Aircraft Impact Assessment Rule Consideration of Aircraft Impacts for New Nuclear Power Reactors," 06/12/2009

4. Draft Regulatory Guide DG 1176, "Guidance for the Assessment of Beyond-Design-Basis Aircraft Impacts," 07/31/2009 (ML073170252)

5. NEI 07-13, Revision 7, "Methodology for Performing Aircraft Impact Assessments for New Plant Designs," 05/31/2009 (ML093570239)

6. Inspection procedure IP 37804, "Aircraft Impact Assessment," 04/27/2010 (ML100680701)

7. Letter to Chairman Jaczko, Draft Final Regulatory Guide 1.217, "Guidance for the Assessment of Beyond-Design-Basis Aircraft Impacts," 02/18/2010 (ML100470861)

Distribution:
See next page

**Accession No:** ML102630190    **Publicly Available (Y/N):**Y        **Sensitive (Y/N):**N
If Sensitive, which category?
**Viewing Rights:** ☐ NRC Users  or  ☐ ACRS only  or  ☐ See restricted distribution

| OFFICE | ACRS | ACRS | ACRS | ACRS | NRO |
|--------|------|------|------|------|-----|
| NAME | MBanerjee | CSantos | EHackett | EHackett for SAbdel-Khalik | NGilles/SUNSI Review Only |
| DATE | 9/20/10 | 9/20/10 | 9/20/10 | 9/20/10 | 9/23/10 |

**OFFICIAL RECORD COPY**

Letter to the Honorable Gregory B Jaczko, Chairman, NRC, from Said Abdel-Khalik, Chairman, ACRS, dated September 20, 2010

SUBJECT: REPORT ON THE SAFETY ASPECTS OF THE SOUTH TEXAS PROJECT NUCLEAR OPERATING COMPANY APPLICATION TO AMEND THE CERTIFIED U.S. ABWR DESIGN TO INCORPORATE THE AIRCRAFT IMPACT ASSESSMENT RULE

Distribution:
ACRS Staff
ACRS Members
B. Champ
A. Bates
S. McKelvin
L. Mike
J. Ridgely
N. Gilles
M. Tonacci
S. Joseph
RidsSECYMailCenter
RidsEDOMailCenter
RidsNMSSOD
RidsNSIROD
RidsFSMEOD
RidsRESOD
RidsOIGMailCenter
RidsOGCMailCenter
RidsOCAAMailCenter
RidsOCAMailCenter
RidsNRROD
RidsNROOD
RidsOPAMail
RidsRGN1MailCenter
RidsRGN2MailCenter
RidsRGN3MailCenter
RidsRGN4MailCenter

# APPENDIX F
# STAFF RESPONSE TO ADVISORY COMMITTEE ON REACTOR SAFEGUARDS

███████████████████████

October 27, 2010

Dr. Said Abdel-Khalik, Chairman
Advisory Committee on Reactor Safeguards
U.S. Nuclear Regulatory Commission
Washington, DC  20555-0001

SUBJECT: RESPONSE TO THE ADVISORY COMMITTEE ON REACTOR SAFEGUARDS
LETTER:  REPORT ON THE SAFETY ASPECTS OF THE SOUTH TEXAS
PROJECT NUCLEAR OPERATING COMPANY'S APPLICATION TO AMEND THE
CERTIFIED U.S. ADVANCED BOILING-WATER REACTOR DESIGN TO
INCORPORATE THE AIRCRAFT IMPACT ASSESSMENT RULE

Dear Dr. Abdel-Khalik:

I am writing in response to the letter from the Advisory Committee on Reactor Safeguards
(ACRS), dated September 20, 2010.  The letter addresses the U.S. Nuclear Regulatory
Commission (NRC) staff's safety evaluation (SE) of the South Texas Project Nuclear Operating
Company's (STPNOC's) application to amend the certified U.S. Advanced Boiling-Water
Reactor (ABWR) design.  ACRS discussed the SE during its 575th meeting on
September 9–11, 2010.  The ACRS ABWR subcommittee had met previously on
August 18, 2010, to discuss the technical aspects of the application, the aircraft impact
assessment (AIA), and the staff's SE and AIA inspection report.

In its letter, the ACRS stated that the STPNOC application to amend the ABWR design
certification rule and the staff's SE are acceptable upon satisfactory closure of
Recommendations 1 and 2 only.  The staff's responses documenting the satisfactory closure of
Recommendations 1 and 2 are given below.  In addition, the ACRS letter contained three
additional recommendations to the staff.  The enclosure to this letter details with the staff's
responses to those three additional ACRS recommendations.

> ACRS Recommendation 1:  The STPNOC application to amend the ABWR
> design certification rule and the staff's SE are acceptable subject to satisfactory
> closure of the issues identified in the Notice of Violation and Recommendation 2.

Staff Response:  Based on the applicant's initial response to the notice of violation (Agencywide
Documents Access and Management System (ADAMS) Accession No. ML102590073) and its
subsequent letter containing additional clarification on two items (ADAMS Accession No.
ML102850361), the staff has accepted the applicant's response to the notice of violation
(ADAMS Accession No. ML102861857) and the staff considers this issue closed.

███████████████████████

S. Abdel-Khalik　　　　　　　　　　　　- 2 -

ACRS Recommendation 2:  The staff should ensure that the applicant demonstrates and documents that the temperature within the fire-protected area where the Alternate Feedwater Injection (AFI) system instrument rack is to be located will not exceed the instruments' environmental qualification conditions.

Staff Response:  The applicant completed the calculations demonstrating that the temperature within the fire-protected area in which the AFI system instrumentation rack will be located will not rise to a level that exceeds the environmental qualification of the instrumentation.  The applicant included those calculations as an appendix to the AIA heat removal report.  The staff inspected the appendix and confirmed that the temperature limits of the AFI instrumentation will not be exceeded and that the calculation is documented in the AIA.

The staff appreciates the Committee's efforts and recommendations.  We thank the ACRS for its time and its valuable input, and we look forward to working with the Committee in the future.

Sincerely,

*/RA by Martin J. Virgilio for/*

R. W. Borchardt
Executive Director
 for Operations

Docket No.:  52-001

Enclosure:
As stated

cc:  Chairman Jaczko
　　　Commissioner Svinicki
　　　Commissioner Apostolakis
　　　Commissioner Magwood
　　　Commissioner Ostendorff
　　　SECY

ACRS Recommendation 2: The staff should ensure that the applicant demonstrates and documents that the temperature within the fire-protected area where the Alternate Feedwater Injection (AFI) system instrument rack is to be located will not exceed the instruments' environmental qualification conditions.

Staff Response: The applicant completed the calculations demonstrating that the temperature within the fire-protected area in which the AFI system instrumentation rack will be located will not rise to a level that exceeds the environmental qualification of the instrumentation. The applicant included those calculations as an appendix to the AIA heat removal report. The staff inspected the appendix and confirmed that the temperature limits of the AFI instrumentation will not be exceeded and that the calculation is documented in the AIA.

The staff appreciates the Committee's efforts and recommendations. We thank the ACRS for its time and its valuable input, and we look forward to working with the Committee in the future.

Sincerely,

*/RA by Martin J. Virgilio for/*

R. W. Borchardt
Executive Director
 for Operations

Docket No.: 52-001

Enclosure:
As stated

cc: Chairman Jaczko
    Commissioner Svinicki
    Commissioner Apostolakis
    Commissioner Magwood
    Commissioner Ostendorff
    SECY

DISTRIBUTION: G20100608/EDATS: SECY-2010-0467/LTR-10-0440
See next page

| ADAMS Accession No.: ML102800594 (Package) | | *via e-mail | EDO-002 |
|---|---|---|---|
| OFFICE | NRO/DNRL/BWR/PM | NRO/DNRL/NRGA/PM | NRO/DNRL/BWR/LA | Tech Editor* |
| NAME | SJoseph | NGilles | BAbeywickrama | KAzariah-Kribbs |
| DATE | 10/15/10 | 10/18/10 | 10/15/10 | 10/18/10 |
| OFFICE | NRO/DE/SEB1 | NRO/DCIP/CQVA | NRO/DNRL/BWR/BC | OGC |
| NAME | BThomas | JPeralta | MTonacci | MSpencer (SKirkwood for) |
| DATE | 10/18/10 | 10/18/10 | 10/19/10 | 10/21/10 |
| OFFICE | NRO/DNRL/DD | NRO/OD | EDO | |
| NAME | DMatthews (FAkstulewicz for) | MJohnson | RWBorchardt (MVirgilio for) | |
| DATE | 10/21/10 | 10/25/10 | 10/27/10 | |

OFFICIAL RECORD COPY

Letter to Dr. Said Abdel-Khalik from R. W. Borchardt dated October 27, 2010

SUBJECT:  RESPONSE TO THE ADVISORY COMMITTEE ON REACTOR SAFEGUARDS
          LETTER:  REPORT ON THE SAFETY ASPECTS OF THE SOUTH TEXAS
          PROJECT NUCLEAR OPERATING COMPANY'S APPLICATION TO AMEND THE
          CERTIFIED U.S. ADVANCED BOILING-WATER REACTOR DESIGN TO
          INCORPORATE THE AIRCRAFT IMPACT ASSESSMENT RULE

**DISTRIBUTION:  G20100608/EDATS: SECY-2010-0467/LTR-10-0440**
Docket No. 52-001
PUBLIC
BWR R/F
MTonacci, NRO
SJoseph, NRO
BAbeywickrama, NRO
NGilles, NRO
MCaruso, NRO
GWunder, NRO
DAndrukat, NRO
JDreisbach, NRO
DJeng, NRO
BJain, NRO
RidsNroOd
RidsNroMailCenter
RidsNroDnrl
RidsNroDnrlNrga
RidsNroDnrlBwr
RidsOgcMailCenter
RidsEdoMailCenter
RidsAcrsAcnwMailCenter

**STAFF RESPONSE TO ADVISORY COMMITTEE ON REACTOR SAFEGUARDS RECOMMENDATIONS 3, 4, AND 5 ON THE SOUTH TEXAS PROJECT NUCLEAR OPERATING COMPANY'S APPLICATION TO AMEND THE CERTIFIED U.S. ADVANCED BOILING-WATER REACTOR DESIGN TO INCORPORATE THE AIRCRAFT IMPACT ASSESSMENT RULE**

In addition to the two recommendations addressed within the staff's letter, the ACRS identified three additional recommendations to the staff. The staff responses to the additional ACRS recommendations are given below.

> ACRS Recommendation 3: The staff should ensure that the assumptions and initial conditions credited in the applicant's Aircraft Impact Assessment (AIA) are properly incorporated into the amended Design Control Document (DCD).

Staff Response: The staff agrees with this comment. Following the inspection of the applicant's AIA, the staff ensured that the amended DCD included the key assumptions related to the credited design features and functional capabilities necessary to ensure the plant is constructed in accordance with the AIA rule (Title 10 of the *Code of Federal Regulations* (10 CFR) Part 50, Section 50.150, "Aircraft Impact Assessment"). The AIA rule does not require applicants to document all their assumptions and initial conditions in the DCD. The staff believes that those items that it required the applicant to document in the DCD are the key assumptions and initial conditions required for inclusion by the AIA rule. The level of detail with regard to assumptions and initial conditions that South Texas Project Nuclear Operating Company (STPNOC) included in its amended DCD is consistent with the level of detail normally provided in DCDs.

The ACRS letter cited the example of the AIA assumption of divisional separation for cable routing. Section 9.5.1.1.1, "Plant Arrangement", of the DCD states that the plant arrangement for the ABWR naturally groups cable trays together in divisional arrangements and does not require routing of services of one division across space allotted to another division. The term "space" refers to physical space and not just fire area. Section 9.5.1.1.2, "Divisional Separation", describes that in general, systems are grouped together by safety division so that there is only one division of safe shutdown equipment in a fire area, but that there are areas where there is equipment from more than one safety division in a fire area. Section 9A.5.5, "Fire Separation for Divisional Electrical Systems," clearly describes each case where cables of more than one division are in relatively close proximity and require special justification. The AIA accounted for the special cases detailed in Section 9A.5.5 of the DCD. The key assumptions made in the AIA are consistent with the descriptions for divisional separation of cables in the ABWR DCD. In addition, the applicant has documented in the DCD that for the alternate feedwater injection (AFI) system, a credited system that survives all impact scenarios, the required cabling will be routed outside of physical damage footprints.

Enclosure

- 2 -

ACRS Recommendation 4: The staff should ensure that Combined License (COL) applicants referencing this amendment have an appropriate process to assure the reliability of the AFI system.

Staff Response: COL applicants will establish the structure, systems, components (SSCs) in the plant-specific maintenance program and other programs that contribute to reliability assurance. This will be accomplished by using the advanced boiling-water reactor design reliability assurance and maintenance programs described in the DCD as starting points and augmenting them as necessary based on the plant-specific design and the plant-specific Probabilistic Risk Assessment (PRA). This activity will occur after the applicant has received a COL. If the COL applicant decides to credit the AFI system in the PRA for providing core cooling for beyond-design-basis accidents, then the AFI system would be considered in the scope of the reliability assurance and maintenance programs. In that case, the NRC would use its reactor oversight process to ensure that these programs address the SSCs associated with the AFI system commensurate with their importance to risk. The DCD amendment has not credited the AFI system for mitigation of non-AIA design basis events and therefore, it is not considered, nor is it required, to be in the scope of the reliability assurance and maintenance programs. The AIA rule does not require applicants to include systems credited for mitigation of this beyond-design-basis accident in the scope of the reliability assurance and maintenance programs. At a minimum, and as stated in Section 9.5.14.3 of STPNOC's amended DCD, licensees that have referenced this DCD will test and maintain SSCs associated with the AFI in accordance with the manufacturer's recommendations to ensure the reliability of the AFI system.

ACRS Recommendation 5: The staff should complete a lessons-learned review of this application to identify any deficiencies in the AIA Inspection Procedure (IP) 37804 and the AIA methodology prescribed in Nuclear Energy Institute (NEI) 07-13, Revision 7. Resolution of these deficiencies should be communicated to the industry and incorporated into the staff's future reviews.

Staff Response: The staff agrees with the ACRS recommendation to complete a lessons-learned review of this application. The staff is currently in the process of formally documenting the lessons-learned as a result of the initial AIA inspections. As the staff completes each AIA inspection, it is informally incorporating lessons-learned into the inspections for each subsequent design center application. The staff also plans to communicate its lessons-learned to the industry and to update the AIA guidance documents and inspection procedure by the end of the third quarter of fiscal year 2011.

F-6

| NRC FORM 335<br>(12-2010)<br>NRCMD 3.7 | U.S. NUCLEAR REGULATORY COMMISSION | 1. REPORT NUMBER<br>(Assigned by NRC, Add Vol., Supp., Rev.,<br>and Addendum Numbers, if any.) |
|---|---|---|
| | **BIBLIOGRAPHIC DATA SHEET**<br>*(See instructions on the reverse)* | NUREG - 1948 |

| 2. TITLE AND SUBTITLE | 3. DATE REPORT PUBLISHED | |
|---|---|---|
| Final Safety Evaluation Report Related to the Aircraft Impact Amendment to the U.S. Advanced Boiling Water Reactor (ABWR) Design Certification | MONTH | YEAR |
| | June | 2011 |
| | 4. FIN OR GRANT NUMBER | |

| 5. AUTHOR(S) | 6. TYPE OF REPORT |
|---|---|
| | Final Safety Evaluation |
| | 7. PERIOD COVERED *(Inclusive Dates)* |

8. PERFORMING ORGANIZATION - NAME AND ADDRESS *(If NRC, provide Division, Office or Region, U.S. Nuclear Regulatory Commission, and mailing address; if contractor, provide name and mailing address.)*

Division of New Reactor Licensing, Office of New Reactors, U.S. Nuclear Regulatory Commission
Washington, DC 20555-0001

9. SPONSORING ORGANIZATION - NAME AND ADDRESS *(If NRC, type "Same as above"; if contractor, provide NRC Division, Office or Region, U.S. Nuclear Regulatory Commission, and mailing address.)*

Same as above

10. SUPPLEMENTARY NOTES
Docket No. 52-0001

11. ABSTRACT *(200 words or less)*

This Safety Evaluation Report (SER) documents the technical review of the U.S. Advanced Boiling Water Reactor (ABWR) aircraft impact assessment (AIA) application by the U.S. Nuclear Regulatory Commission (NRC) staff. The STP Nuclear Operating Company (STPNOC) submitted the application on June 30, 2009 to amend the design certification rule for the ABWR in accorance with the procedures of Subpart B, "Standard Design Certifications" of Title 10 of the Code of Federal Regulations (10 CFR) Part 52, "Licenses, Certifications, and Approvals for Nuclear Power Plants". On September 23, 2010, STPNOC submitted the final amendment application (Agencywide Documents Access and Management System Accession No. ML102870017). The purpose of the amendment is to address the requirements in 10 CFR 50.150, "Aircraft Impact Assessment". If the proposed amendment is approved, applicants for a combined license that reference the ABWR standard design may address the requirements of 10 CFR 50.150 by referencing the amended ABWR standard design. On the basis of its evaluation and independent analyses, the NRC staff concludes that STPNOC's application meets the requirements of Subpart B of 10 CFR Part 52 and 10 CFR 50.150(b).

| 12. KEY WORDS/DESCRIPTORS *(List words or phrases that will assist researchers in locating the report.)* | 13. AVAILABILITY STATEMENT |
|---|---|
| Aircraft Impact Assessment (AIA)<br>South Texas Project Nuclear Operating Company<br>Advanced Boiling Water Reactor (ABWR)<br>10 CFR 50.150 | unlimited |
| | 14. SECURITY CLASSIFICATION |
| | *(This Page)* |
| | unclassified |
| | *(This Report)* |
| | unclassified |
| | 15. NUMBER OF PAGES |
| | 16. PRICE |

UNITED STATES
**NUCLEAR REGULATORY COMMISSION**
WASHINGTON, DC 20555-0001

OFFICIAL BUSINESS

NUREG-1948

Final Safety Evaluation Report Related to the Aircraft Impact Amendment to the U.S. Advanced Boiling Water Reactor (ABWR) Design Certification

June 2011

www.ingramcontent.com/pod-product-compliance
Lightning Source LLC
Chambersburg PA
CBHW080302180526
45167CB00006B/2642